Project-Based R Companion to Introductory Statistics

Project-Based R Companion to Introductory Statistics

Chelsea Myers

CRC Press
Taylor & Francis Group
Boca Raton London New York

CRC Press is an imprint of the
Taylor & Francis Group, an **informa** business
A CHAPMAN & HALL BOOK

First edition published 2021
by CRC Press
6000 Broken Sound Parkway NW, Suite 300, Boca Raton, FL 33487-2742

and by CRC Press
2 Park Square, Milton Park, Abingdon, Oxon, OX14 4RN

Library of Congress Cataloging-in-Publication Data

Names: Myers, Chelsea, author.
Title: Project-based R companion to introductory statistics / Chelsea Myers.
Description: First edition. | Boca Raton : Taylor and Francis, 2021. | Includes bibliographical references and index.
Identifiers: LCCN 2020041189 (print) | LCCN 2020041190 (ebook) | ISBN 9780367262082 (paperback) | ISBN 9780429292002 (ebook)
Subjects: LCSH: Mathematical statistics—Data processing. | R (Computer program language)
Classification: LCC QA276.4 .M94 2021 (print) | LCC QA276.4 (ebook) | DDC 519.50285/5133—dc23
LC record available at https://lccn.loc.gov/2020041189
LC ebook record available at https://lccn.loc.gov/2020041190

ISBN: 978-0-367-68734-2 (hbk)
ISBN: 978-0-367-26208-2 (pbk)
ISBN: 978-0-429-29200-2 (ebk)

Typeset in Minion Pro
by KnowledgeWorks Global Ltd.

Access the Support Material: https://www.routledge.com/Project-Based-R-Companion-to-Introductory-Statistics/Myers/p/book/9780367262082

Contents

Author

After a 10-year career as a research biostatistician in the Department of Ophthalmology and Visual Sciences at the University of Wisconsin-Madison, Chelsea Myers teaches statistics and biostatistics at Rollins College and Valencia College in Central Florida. She has authored or coauthored more than 30 scientific papers and presentations and is the creator of the MCAT preparation website MCATMath.com. She lives in Winter Park, Florida with her family.

Introduction

One of the most difficult transitions students face when beginning a first "real" job after college is moving beyond solving the well-designed problems found in textbooks to tackling the messy, complex problems they face in the real world. Textbooks and well-designed problems are an important part of a course curriculum. In fact, this book is far less valuable to students without a companion statistics textbook that can develop the mathematical concepts needed to accurately interpret results of statistical tests. And engaging in multiple well-thought-out practice exercises covering the same topic allows students to work toward mastery of a subject.

However, interesting scientific questions and thorny business issues don't come labeled with neat chapter headings explaining exactly what methods are used to solve the problem. Instead, they come as large, multistep projects with a complex decision-making process and lots of room for errors – those caused by humans and those beyond anyone's control. That is the niche this book sits in.

Our aim is to guide students away from the neat and tidy world of textbook questions and toward an accurate portrayal of real-world data analysis. To that end, all of the datasets used in this book come from actual analyses. Some are large, some are small, some may seem confusing at first glance, some have errors, but all were collected with the goal of asking and answering important research questions.

Each chapter includes at least one worked example called a Guided Project, which is followed by a short description of the dataset and set of deliverables. At least one unworked Student Project will follow the Guided Project and will have its own short description of the dataset and list of deliverables.

A NOTE TO STUDENTS

If you have any previous experience working with software that can do statistical analyses, it's likely using Microsoft Excel. RStudio has many features that are point-and-click like Excel, but the commands used to run RStudio look a bit like lines of computer code, and that can be intimidating to some.

However, we encourage you to put aside any preconceived notions about computer programming and instead focus on a different maxim: Working with RStudio isn't *harder* than Excel (or another software you've used before), it's just *different*.

A NOTE TO INSTRUCTORS

This book is organized in chapters that should dovetail nicely with those in most introductory statistics textbooks and is intended as a companion to a traditional course text. It will, by necessity, leave out many technical and mathematical details found in the primary textbook.

Instead, each chapter will present its material using a complete step-by-step analysis of a real, publicly available dataset with an emphasis on the practical skills of testing assumptions, data exploration, and forming conclusions.

The worked examples in each chapter are titled Guided Projects, and each is followed by a short description of the dataset and set of deliverables. Each deliverable is a specific question that should be answered as part of the larger project. A Student Project will follow the Guided Projects in each chapter and will have its own short description of the dataset and list of deliverables.

Guided Projects are excellent material for summary lectures at the end of a topic or for demonstration in a lab section. Student Projects are useful as take-home or laboratory assignments. At the back of the book there are several summative Student Projects that incorporate skills from multiple chapters and would make good final projects or take-home final exams. An instructor's solutions manual for all Student Projects is available.

Access support material at https://www.routledge.com/Project-Based-R-Companion-to-Introductory-Statistics/Myers/p/book/9780367262082

CHAPTER 1

Getting Started with R and RStudio

WORKING WITH RStudio

The first step in using RStudio is to download it and install it on your personal computer. R and RStudio are free to use and easy to install on Mac and Windows computers. It's a bit more complicated to install the software on a Linux or a Unix machine, so if you use one of those operating systems and don't feel comfortable doing the installation yourself, we recommend contacting a systems administrator to guide you through the process.

Note that, while this book focuses on using RStudio, the R software must be installed before RStudio for RStudio to work.

To install R on Windows:

1. Open "http://www.r-project.org" in your browser.
2. Click on "download R" in the paragraph under "Getting Started". You'll see a list of mirror sites organized by country.
3. Click on "https://cloud.r-project.org/" under the heading "0-Cloud".

4. Click on "Download R for Windows" under the heading "Download and Install R".
5. Click on "base".
6. Click on the link for downloading the latest version of R.
7. When the download completes, double-click on the .exe file and follow the prompts.

To install R on OS X:

1. Open "http://www.r-project.org" in your browser.
2. Click on "download R" in the paragraph under "Getting Started". You'll see a list of mirror sites organized by country.
3. Click on "https://cloud.r-project.org/" under the heading "0-Cloud".
4. Click on "Download R for (Mac) OS X" under the heading "Download and Install R".
5. Click on the .pkg file for the most recent version of R under "Latest Release".
6. Click on the link for downloading the latest version of R.
7. When the download completes, double-click on the .pkg file and follow the prompts.

Once you have R installed on your computer, you must install RStudio.

To install RStudio on Windows or Mac:

1. Open "https://www.rstudio.com/" in your browser.
2. Click on "Download".
3. Click on "Download" under "RStudio Desktop Open Source License".

4. Click on the appropriate (Mac or Windows) file under "Installers" to download RStudio.

5. When the download completes, double-click on the file and follow the prompts.

Once RStudio is installed, you can open it just as you would any other application.

GUIDED PROJECT: OPEN AND MODIFY THE `Titanic.xlsx` DATASET IN RStudio

In the early hours of April 15, 1912, the unsinkable ship *RMS Titanic* sank when it struck an iceberg, killing more than half of the passengers and crew aboard. The `Titanic.xlsx` dataset contains demographic information for 889 of those passengers as well as a record of whether or not each passenger survived[1]. Our goal is to explore the functionality of RStudio by opening and modifying the `Titanic` dataset.

Project deliverables:

1. Understand the function of the RStudio Console.

2. Import the `Titanic.xlsx` dataset into RStudio.

3. Use the data dictionary to identify the variables in the dataset.

4. Open and save a new R Script.

5. Create a new variable called `Titanic$Sex.num` that takes on the value `0` if a passenger is `female` and `1` if a passenger is `male`.

6. Create a subset of the `Titanic` dataset called `Titanic.Firstclass` that includes only first-class passengers.

7. Create a subset of the `Titanic` dataset called `Titanic.Children` that includes only passengers under the age of 18.

8. Save, close, and reopen an R Script.

Deliverable 1: Understand the function of the RStudio console. When you open RStudio for the first time, you'll see a screen with three panels. One panel called the Console runs vertically down the left side of the application. There is some text that is preloaded into the Console, and under that text you'll see a > symbol. This is where commands are executed. Click in the window next to the >, type 2+2 and hit the enter/return key (depending on if you are using a PC or Mac). The result should appear in the Console as

```
> 2+2
[1] 4
```

RStudio can do many calculations in the Console; however, more often, we want to open a dataset that has already been created. In this text, we will focus on opening data stored as an Excel (.xlsx) file. You will need to download each dataset used in the Guided and Student Projects and save them on your computer where you can find the files easily.

Deliverable 2: Import the `Titanic.xlsx` dataset into RStudio. The easiest way to import a file into RStudio is by selecting "File" and then "Import Dataset" from the main menu. When you hover your mouse over "Import Dataset", another menu will appear and allow you to select the type of file you would like to import. In this case, we will choose "From Excel…", but note that you can read in data saved as a text or .csv file or a SAS, Stata or SPSS (which are other statistical software) file, and the process will be largely the same.

Click the "Browse" button to the right of the text box under "File/Url:" and navigate to where you saved the `Titanic.xlsx` file. Click on `Titanic.xlsx` and "Open" to begin the process of importing the dataset into RStudio. It will take a few seconds, but if the import worked correctly, the panel labeled "Data Preview" will be populated with the column names and the first several rows of data from the Excel file. It should look very much like the original Excel spreadsheet.

Note the panel labeled "Code Preview" at the bottom left. This panel contains the code RStudio will automatically paste in the Console to import that data after you click the "Import" button.

Click "Import". You will return to the original screen, but note that several things have changed. First, the left pane of the screen has now split in two with a spreadsheet view of the `Titanic` dataset at the top and the Console on the bottom. The commands in the Console now read

```
> library(readxl)
> Titanic <- read_excel("~[Your directory path]/
  Titanic.xlsx")
> View(Titanic)
```

The > `View(Titanic)` command inserted automatically by RStudio is what prompted the spreadsheet to appear above the Console.

On the upper right hand side of the screen under "Environment", there is a list of datasets that have been created or uploaded during the session. Now `Titanic` appears in that list as well as details about the number of observations and variables in the dataset. Double clicking on `Titanic` under "Environment" has the same effect as entering `View(Titanic)` in the Console.

If you import additional datasets, they will also be listed under "Environment".

Deliverable 3: Use the data dictionary to identify the variables in the dataset and the meaning of any codes used for variables

Data sets often contain abbreviations for variables and use codes to represent different values of the variable. The data dictionary is where all of these abbreviations and codes are explained. You will save yourself time and heartache if you make it a habit to always consult the data dictionary as you explore your dataset. What seems intuitive often isn't. There is a companion data dictionary for every dataset used in this text.

Open the file "*Titanic* Data Dictionary". We recommend saving it to your local directory in the same place you've saved the `Titanic.xlsx` dataset.

Each data dictionary used in this book has three columns. The first lists the variable name as it appears in the Excel file. The second gives a longer description of the variable. This is particularly helpful when the variable names are short. For example, it might not be obvious when you first open the `Titanic` dataset that `Pclass` refers to the passenger's ticket class. The variable description gives you that information.

Finally, the "Details" column tells you the values you can expect to see for each variable in the dataset. The first piece of information under "Details" tells you the data type of the variable: the "character" type is a sequence of one or more text characters, like a word or phrase, and the "numeric" type represents a decimal number. Then in brackets you see the values the variable can take on. Some variables can only take on a few values. For example, `Survived` can either equal `0` or `1`. `Sex` is either `male` or `female`. `Pclass` is equal to `1`, `2`, or `3`. Some variables can take on an almost infinite number of values such as `Name`, which can be any string of characters. For numeric variables that are measurements, the brackets will identify the units of measure. For example, `Age` is measured in `years` and `Fare` is measured in `£`.

After the brackets, there is additional information to help the user interpret what each value of the variable means. While `Sex = male` and `Sex = female` may seem clear from the context, `Survived = 0` and `Survived = 1` is not. We must consult the data dictionary to know that `Survived = 0` means the passenger did not survive and `Survived = 1` means he or she did. Many datasets use this 0 = No, 1 = Yes convention, but never take it for granted. Also, make sure you are very clear on what a variable represents. If the variable `Survived` had instead been named `Perished`, the 0 = No, 1 = Yes convention could still hold but would mean exactly the opposite!

The first variable in the `Titanic` dataset is `Survived`. This variable indicates if the passenger survived the *Titanic* sinking or not. It is coded `0` if the passenger did not survive and `1` if the passenger survived. The second variable is `Pclass`. This identifies the passenger's ticket class. It is coded `1` if the passenger was in first class, `2` if the passenger was in second class, and `3` if the passenger was in third class. The third variable is `Name`, which records the name of each passenger as a character string. The fourth variable is `Sex`, which identifies the passenger as `male` or `female` in a character string. The fifth variable is `Age`, which records the passenger's age in years. The sixth variable is `Siblings/Spouse_Aboard`, which records the total number of individuals related to the passenger as a sibling or spouse who were also onboard. The seventh variable is `Parents/Children_Aboard`, which records the total number of individuals related to the passenger as a parent or child who were also onboard. Finally, the last variable is `Fare`, which records the amount paid (in £) for the passenger's ticket.

Deliverable 4: Open and save a new R Script.

In Deliverable 1 we learned that RStudio works by executing commands that are typed in the Console window. However, it's important to know that nothing that you type in the Console will be saved when you close RStudio (even if you save your Workspace!). A better way to save and rerun R code is by writing an R Script. Scripts can be saved, reopened, and run over and over again.

To open a new script, select "File", then "New File", and "R Script". This will open a new tab in the window above the Console where we previously viewed the `Titanic` data set in spreadsheet form. Type 2+2 on line 1 and press enter. This time 2+2 is not executed, and the cursor just moves down to the next line.

To run the code, highlight it with the mouse, and click "Run" in the menu bar of the tab. This will paste 2+2 in the Console and execute the code. The results should look exactly the same as when we entered 2+2 in the Console directly in Deliverable 1.

```
> 2+2
[1] 4
```

To save the R Script, click your cursor in the R Script tab. Select "File" then "Save As…" Navigate to the location where you want to save the R Script and name it. It will be saved as a .R file. You can continue to edit and resave the script periodically as you would any other document.

Deliverable 5: Create a new variable called `Titanic$Sex.num` that takes on the value `0` if a passenger is `female` and `1` if a passenger is `male`

One of the most common data manipulations we'll need to do is to create a new variable and add it to a dataset. In the `Titanic` dataset, `Sex` is recorded as `male` or `female`. However, it is sometimes more useful to represent data using a numeric code rather than a character string (or vice versa). Therefore, we'd like to create a new variable called `Sex.num` in the `Titanic` dataset that is coded as `0` if the passenger was `female` and `1` if the passenger was `male`.

To do this, we will use the assignment symbol (<-) and the `ifelse` function. The assignment symbol assigns data to the new variable name. To create the `Sex.num` variable in the `Titanic` dataset, type the following in your R Script:

```
Titanic$Sex.num <-
     ifelse(
          test = (Titanic$Sex == 'female'),
          yes = 0,
          no = 1
            )
```

The `$` between `Titanic` and `Sex.num` specifies that `Sex.num` is a variable that is to be created in the `Titanic` dataset. The assignment symbol assigns new data to `Sex.num` based on the results of the `ifelse` function. Note that dataset and variable names can't have any spaces so we sometimes choose to use an underscore or period if we'd like to assign a name that has more than one word.

The `ifelse` function takes three input arguments named `test`, `yes`, and `no`. The values assigned to each of these three arguments control the behavior of the function: `test` is a logical expression that evaluates to `true` or `false`, `yes` is the value to return if the

TABLE 1.1 Row Added to the `Titanic.xlsx` Dataset

Variable Name	Description	Details
Sex.num	Passenger sex	Numeric [0,1] 0 = female, 1 = male

expression is `true`, and `no` is the value to return if the expression is `false`. Notice how the = symbol is used to assign values to the three arguments.

Within the `ifelse` function, we see the logical statement `Sex == 'female'`. In R syntax, the double equal sign means "is exactly equal to". The `ifelse` function looks at the variable `Sex` for every passenger in the dataset. When `Sex` is `female` (the logical expression `Sex == 'female'` is `true`), then `Titanic$Sex.num` is assigned the value `0`. When `Sex` is `male` (the logical expression `Sex == 'female'` is `false`), then `Titanic$Sex.num` is assigned the value `1`.

Once you have typed the code for `Titanic$Sex.num` in your R Script, highlight all the rows and click "Run". Then type and run `View(Titanic)` in your R Script to look at the updated spreadsheet view of the `Titanic` dataset. Check to see that the new `Sex.num` variable takes on the value `1` when the passenger is `male` and `0` when the passenger is `female`.

It is also good practice to add this variable to the data dictionary in case someone comes along after you and uses your dataset. The inserted row appears in Table 1.1:

Deliverable 6: Create a subset of the `Titanic` dataset called `Titanic.Firstclass` that includes only first-class passengers. A second helpful function when working with datasets is `subset`. The R command that will create a new dataset called `Titanic.Firstclass` is

```
Titanic.Firstclass <-
      subset(
            x = Titanic,
            subset = (Titanic$Pclass == 1)
             )
```

The `subset` function takes two input arguments called `x` and `subset`. Similar to the `ifelse` function, the values assigned to these two arguments control the behavior of the function: `x` is the complete dataset you wish to subset, and `subset` specifies which records you wish to include in the subset. In this case, we wish to subset the complete `Titanic` dataset and include records where the passenger class is exactly equal to `1` (`Pclass == 1`).

Type the command to create the `Titanic.Firstclass` dataset in your R Script and run the code. Then type and run `View(Titanic.Firstclass)`. The new data set will appear in spreadsheet view, and you can confirm that it only contains records from first class passengers.

Deliverable 7: Create a subset of the `Titanic` dataset called `Titanic.Children` that includes only passengers under the age of 18.

Similarly, we can create a subset that only includes records for the children who were onboard. The first entry in the `subset` function is still `Titanic`, but now we modify the `subset =` entry to read `Titanic$Age < 18`.

```
Titanic.Children <-
     subset(
          x = Titanic,
          subset = (Titanic$Age < 18)
          )
```

In your R Script, type and run the code to create the `Titanic.Children` dataset. Note that both `Titanic.Firstclass` and `Titanic.Children` now appear under "Data" in the Environment window in the upper right panel of RStudio. Double-click on each dataset name in that panel to see them appear in spreadsheet view on the upper left.

Deliverable 8: Save, close, and reopen an R Script.

It would be cumbersome to have to go through the process of importing each dataset you want to work with each time you open RStudio.

Fortunately, by using and saving R Scripts, you can preserve your work from session to session.

First, find the section of code that appeared in the Console when you imported the `Titanic` dataset. It will look something like this:

```
> library(readxl)
> Titanic <- read_excel("~[Your directory path]/
  Titanic.xlsx")
> View(Titanic)
```

Copy and paste those lines of code at the top of your R Script. Running this code when you open RStudio will allow you to import `Titanic.xlsx` without using menus like we did in Deliverable 2. Make sure to save your R Script.

When you close and reopen RStudio, it will, by default, reload your previous workspace (even if you didn't save it!). This might strike some people as convenient and others as bizarre, but it is a feature we want to turn off so we can start with a blank slate every time we open RStudio. Otherwise, it is easy to accidentally use an old version of a dataset or code, which could cause reproducibility errors.

To disable this automatic reloading feature, select "Tools" and then "Global Options" from the main menu. Uncheck the boxes for "Restore most recently updated project at startup", "Restore previously open source documents at startup", "Restore .RData into workspace at startup", and "Always save history (even when not saving .RData)". Finally, in the drop down menu, change "Save workspace to .RData on Exit" to "Never". Then click "Apply" and "OK".

Now (after being sure to save your R Script!), quit RStudio by closing the application. Reopen RStudio and it should look just as it did when you opened it for the first time. Select, "File", then "Open File", and select the R Script you wrote for this project. When the script opens, rerun the lines of code. The `Titanic` dataset will be reloaded, and any other functions you saved in the script (creating the `Titanic.children` and `Titanic.Firstclass` subsets, for example) will be generated.

STUDENT PROJECT: OPEN AND MODIFY THE `Ohio.xlsx` DATASET IN RStudio

Approximately 618,000 Union and Confederate soldiers died in battle and from starvation and disease during the single bloodiest conflict in United States history: The Civil War. To put this in perspective, "only" about 400,000 US soldiers perished during the second-most-deadly conflict, World War II. Despite the terrible conditions soldiers faced in the field, remarkably good records about these men, their backgrounds, and their fates have been preserved.

The `Ohio.xlsx` dataset contains records from 45 companies from Ohio during the United States Civil War including the number of soldiers in the company, the year the company was formed, details about the demographic makeup of each company as well as the mortality rate due to injury and illness[2]. Our goal is to continue to explore the functionality of RStudio by opening and modifying the `Ohio.xlsx` dataset.

Project deliverables:

1. Save the `Ohio.xlsx` dataset to a local drive on your computer.
2. Import the dataset into RStudio.
3. Use the data dictionary to identify the variables in the dataset.
4. Open and save a new R Script.
5. Create a new variable called `Ohio$Majority.illness` that takes on the value `1` if the number of soldiers who died due to illness is greater than the number of soldiers who died due to injury (`Death_illness > Death_injury`) and takes on the value `0` otherwise.
6. Add information for the newly created `Majority.illness` variable to your saved copy of the data dictionary.
7. Create a subset of the `Ohio` dataset that contains only companies with enlistment in 1861. Name that new dataset `Ohio.1861`.

REFERENCES

1. Stanford University. (2016). A *Titanic* Probability. *Titanic* Training Dataset. Accessed May 14, 2109. https://web.stanford.edu/class/archive/cs/cs109/cs109.1166/problem12.html
2. Lee, C. (1999). "Selective Assignment of Military Positions in the Union Army: Implications for the Impact of the Civil War," *Social Science History*, Vol. 23, no. 1, pp. 67–97.

CHAPTER 2

Describing Categorical Data

INTRODUCTION

Most undergraduates first encounter data analysis when they are asked to analyze a dataset that was collected by someone else. This introduction to statistical analysis has both advantages and drawbacks.

It is certainly time efficient. It can take years – even an entire career– to progress from idea to approval to data collection to analysis. Most young researchers don't have the time – either in years or in hours of the day – to design and implement a formal survey, experiment or observational study that will pass peer review.

The main drawback to using someone else's data is that it is not your data. Because you were not involved in the study design, your specific research question might not be answered by the study, and you might have to settle for the answer to a slightly different question that the study was designed to answer.

You also have to approach the data you propose to use with the mind of an archeologist. A study happened in the past and there is a written record of it (the data dictionary, study documents, previously published papers), but it is left for you to understand and interpret what actually happened. If you are lucky (and we recommend this if it is at all possible), you will be able to correspond with someone who was involved in the study design and data collection. Even the best, most thorough data dictionary has an "Oh, but…" lurking somewhere that is only maintained in institutional knowledge.

Although every research question is different, the first step in every data analysis is the same: get to know the data. Calculating basic descriptive statistics allows you to answer some very important questions that will guide you throughout your analyses.

There are – broadly – two types of data that are collected in a statistical study: categorical data and quantitative data. Categorical variables place study subjects into groups (hair color, political party, favorite movie genre, etc.). Quantitative variables measure something (height, weight, number of pets, etc.). Because of the fundamental differences between the two, we devote Chapter 2 to the former and Chapter 3 to the latter. But even though the methods of describing and displaying categorical and quantitative data are different, the goal is largely the same.

CATEGORICAL DATA

The first goal of descriptive statistics is to understand the possible values for each variable, what they mean, and how frequently they occur. There are several ways to accomplish this when we work with categorical data, including making frequency or relative frequency tables, bar charts, and pie charts.

While descriptive statistics are often interesting on their own, it is also important to understand what types of variables are present in a dataset because that will inform the statistical tests you will conduct later on. Within categorical data there are three types of variables: ordinal variables, nominal variables, and identifier variables.

Ordinal data has a natural ordering. For example, the letter grades A, B, C, D, and F. It wouldn't make sense to create a table where the letter grades were ordered – say – C, B, F, A, D.

In contrast, nominal data has no natural ordering. Favorite ice cream flavor – chocolate, vanilla, strawberry, chocolate chip mint, etc. – has no natural ordering. If we were to make a bar chart, we'd have no reason to put chocolate before vanilla or vice versa.

Finally, datasets typically have an identifier variable that is used to name or identify each unique subject in a study. The identifier variable places each record into a group of one and only one and is important for keeping the dataset organized but is not analyzed. Company name in the `Ohio` dataset and passenger name in the `Titanic` dataset both fill the role of identifier variable.

GUIDED PROJECT: DOES THE USE OF ANTISEPTICS DURING SURGERY REDUCE MORTALITY?

Spoiler alert – yes! Though this was actually very controversial at the time. Joseph Lister (yes, like Listerine), an English surgeon in the late 1800s, pioneered the use of carbolic acid as an antiseptic during surgery after observing that it mitigated the smell of sewage waste used to irrigate farm fields with no apparent harm to the livestock grazing there.

Mortality data for individuals who had upper or lower limb amputations before and after the discovery of antiseptics are presented in the `Lister.xlsx` dataset[1].

Project deliverables:

1. Use the data dictionary to identify each variable in the dataset as categorical or quantitative. If the variable is categorical, further identify it as ordinal, nominal, or an identifier variable.

2. Using the `ifelse` function covered in Chapter 1, create a new variable called `Outcome.char` that takes on the value `Did Not Survive` when `Outcome == 0` and `Survived` when `Outcome == 1`. Create a second new variable called `Antiseptic.char` that takes on the value `Before Antiseptics` when `Antiseptic == 0` and `After Antiseptics` when `Antiseptic == 1`.

3. Calculate the frequency and relative frequency of survival for the patients in Lister's study.
4. Calculate the frequency and relative frequency of the time period (before or after the discovery of antiseptics) in which the amputations were performed.
5. Create a bar chart to display the number of patients who survived and did not survive.
6. Create a pie chart to display the percent of patients who had amputations performed before and after the discovery of antiseptics.
7. Create a contingency table to show the joint distribution of survival and antiseptic use.
8. Draw a side-by-side bar chart to show the number of patients who survived and did not survive a limb amputation before and after the discovery of antiseptics.
9. Summarize your findings about the relationship between antiseptic use and survival after an amputation.

Deliverable 1: Use the data dictionary to identify each variable in the dataset as categorical or quantitative. If the variable is categorical, further identify it as ordinal, nominal, or an identifier variable. Using the methods detailed in Chapter 1, import the `Lister.xlsx` dataset and open an R Script. Also open the file "Lister Data Dictionary".

The first variable in the dataset is `ID`, which identifies each unique participant. This is a categorical, identifier variable. The second variable is `Antiseptic`, which indicates if the amputation was performed before or after the discovery of antiseptics. This is a nominal categorical variable. The third variable is `Limb`, which indicates if the patient had a lower or upper limb amputated. This is a nominal categorical variable. The last variable is `Outcome`, which indicates if the patient survived or did not survive after the amputation. This is a nominal categorical variable.

Deliverable 2: Using the `ifelse` function covered in Chapter 1, create a new variable called `Outcome.char` that takes on the value Did Not `Survive` when `Outcome == 0` and `Survived` when `Outcome == 1`. Create a second new variable called `Antiseptic.char` that takes on the value `Before Antiseptics` when `Antiseptic == 0` and `After Antiseptics` when `Antiseptic == 1`.

When creating tables and plots, it is often helpful to work with variables that have descriptive category names rather than 0, 1 codes. Type and run the following code in an R Script to create the `Outcome.char` and `Antiseptic.char` variables (review the Guided Project in Chapter 1 if you need additional help with the `ifelse` function):

```
Lister$Outcome.char <-
     ifelse(
          test = (Lister$Outcome == 1),
          yes = "Survived",
          no = "Did Not Survive"
            )

Lister$Antiseptic.char <-
     ifelse(
          test = (Lister$Antiseptic == 1),
          yes = "After Antiseptics",
          no = "Before Antiseptics"
            )
```

Deliverable 3: Calculate the frequency and relative frequency of survival for the patients in Lister's study.

The `table` function is used to find the frequency (the number of observations) that fall in each category of a categorical variable. We use the statement `table(Lister$Outcome.char)` to calculate the number of passengers who survived and did not survive after a lower or upper limb amputation.

Type and run the statement `table(Lister$Outcome.char)` in your R Script, and the following results will appear in the Console:

```
> table(Lister$Outcome.char)
Did Not Survive         Survived
             22               53
```

We can also apply the `prop.table` function to the `table` function to find the relative frequency (the proportion of observations) that falls in each category. Running the relevant code produces the following output in the Console:

```
> prop.table(table(Lister$Outcome.char))
Did Not Survive         Survived
      0.2933333        0.7066667
```

It can also be useful to multiply the results from the `prop.table` function by 100 using `*100` so the relative frequencies are expressed as percents rather than proportions.

```
> prop.table(table(Lister$Outcome.char)) * 100
Did Not Survive         Survived
       29.33333         70.66667
```

Fifty-three of the patients included in the study (about 71%) survived their amputations and 22 (about 29%) did not.

Deliverable 4: Calculate the frequency and relative frequency of the time period (before or after the discovery of antiseptics) in which the amputations were performed.

Similarly, we use the statement `table(Lister$Antiseptic.char)` to calculate the number of individuals who had amputations performed before and after the discovery of antiseptics. Running the relevant code produces the following output in the Console:

```
> table(Lister$Antiseptic.char)
 After Antiseptics Before Antiseptics
                40                 35
```

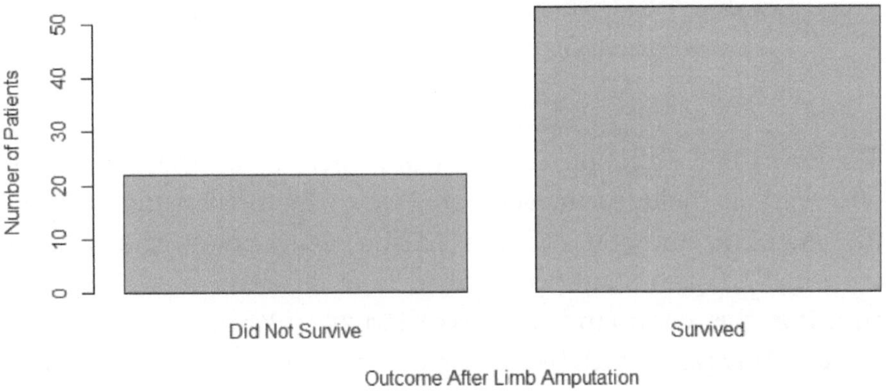

FIGURE 2.1 Patient outcome after limb amputation.

We can use the `prop.table` function and `*100` to find the percent of patients who had amputations performed before and after the discovery of antiseptics.

```
> prop.table(table(Lister$Antiseptic.char)) * 100

  After Antiseptics  Before Antiseptics
           53.33333            46.66667
```

Thirty-five individuals (about 47%) had amputations performed before the discovery of antiseptics, and 40 individuals (about 53%) had amputations performed after the discovery of antiseptics.

Deliverable 5: Create a bar chart to display the number of patients who survived and did not survive.

One nice feature of R is how easy it is to make great looking figures. Below is the R code to create a bar chart displaying the number of patients who survived and did not survive after a limb amputation. We'll walk through each statement in detail after you see Figure 2.1

```
barplot(
      table(Lister$Outcome.char),
      main = "Patient Outcome After Limb Amputation",
```

```
xlab = "Outcome After Limb Amputation",
ylab = "Number of Patients"
 )
```

The `barplot` function creates a bar chart using the number of observations included in each bar. That's why the first statement in the `barplot` function is `table(Lister$Outcome.char)` rather than just `Lister$Outcome.char`. We recommend writing and running the command `barplot(Lister$Outcome.char)` in place of `barplot(table(Lister$Outcome.char))` so you can see the difference.

The `main =` argument specifies the main title for the chart, and the `xlab =` and `ylab =` arguments specify the appropriate labels for the x-axis and- y-axis, respectively.

Deliverable 6: Create a pie chart to display the percent of patients who had amputations performed before and after the discovery of antiseptics.

Recall from Deliverable 3 that the command to calculate the percent of individuals in Lister's study who survived and did not survive a limb amputation was `prop.table(table(Lister$Antiseptic.char)) * 100`. This is the first entry in the `pie` function, which is used to generate a pie chart. As in a bar chart, `main =` is used to give the figure a main title.

We can also specify the colors for the wedges of the pie chart using `col =`. In addition to creating a visually pleasing graph (that might match the color scheme of a poster or PowerPoint presentation), it will allow us to match the legend text to the correct portion of the pie chart. Colors are specified after `col =` using a vector that contains the same number of colors as there are sections of the pie chart. R can produce graphics with an incredible array of colors, a comprehensive list of which can be found here: http://www.stat.columbia.edu/~tzheng/files/Rcolor.pdf. In this example, we will use the colors `darkblue` and `darkcyan (shown in grayscale in the text figure).`

Type and run the following `pie` function in an R Script, which will produce Figure 2.2:

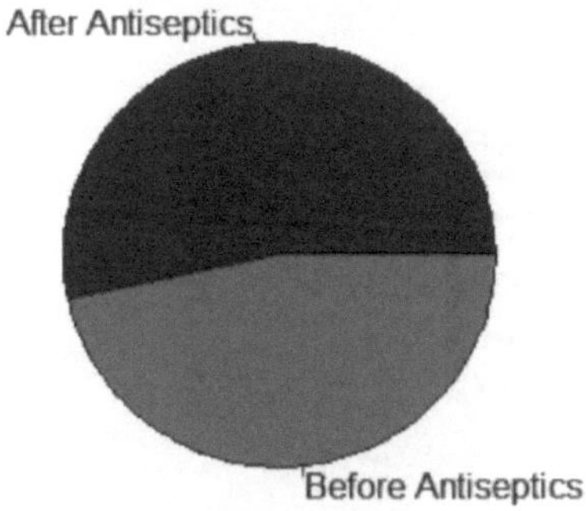

FIGURE 2.2 Percent of patients with amputations performed before and after the discovery of antiseptics.

```
pie(
     prop.table(table(Lister$Antiseptic.char)) * 100,
     main = "Percent of Patients with Amputations
             Performed
             Before and After the Discovery of
             Antiseptics",
     col = c("darkblue", "darkcyan")
   )
```

While the pie chart gives us a general sense of the relative frequency of when individuals had their amputations performed (before or after the discovery of antiseptics), it is helpful to include the actual percent in each category on the graph. We can do this by adding a legend.

The following R code will overlay a legend on the top right of the pie chart we just created:

```
legend(
     x = "topright",
     legend = c("53.3%", "46.7%"),
     fill = c("darkblue", "darkcyan")
     )
```

The first argument, `x = "topright"`, places the legend in the top right corner of the graph. The next argument, `legend =`, specifies

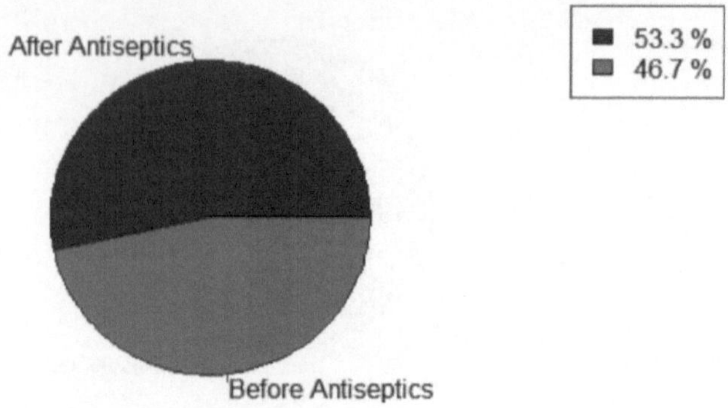

FIGURE 2.3 Percent of patients with amputations perfomed before and after the discovery of antiseptics.

a vector containing the labels to be used in the legend. Looking back at the relative frequency of amputations performed during each time period (which we calculated in Deliverable 3), we see that 53.3% of study participants had an amputation after the discovery of antiseptics and 46.7% had an amputation before the discovery of antiseptics. Finally, we specify the color of the markers in the legend so that they match those in the pie chart command using `fill =` .

The complete R code to generate the pie chart shown in Figure 2.3 is:

```
pie(
     prop.table(table(Lister$Antiseptic.char)) * 100,
     main = "Percent of Patients with Amputations
             Performed
             Before and After the Discovery of
             Antiseptics",
     col = c("darkblue", "darkcyan")
  )
legend(
     "topright",
     legend = c("53.3 %", "46.7 %"),
     fill = c("darkblue", "darkcyan")
     )
```

Note that the `pie` function generates the actual pie chart and the `legend` function overlays the legend on top of the pie chart. If you edit the legend, you'll have to rerun the `pie` function so you have a "clean slate" on which to overlay the revised legend.

Deliverable 7: Create a contingency table to show the joint distribution of survival and antiseptic use.

We can determine the joint distribution of survival and antiseptic use by modifying the `table` function to include both variables of interest separated by a comma. Running the relevant code produces the following output in the Console.

```
> table(Lister$Outcome.char, Lister$Antiseptic.char)

                 After Antiseptics Before Antiseptics
  Did Not Survive                6                 16
  Survived                      34                 19
```

Before the discovery of antiseptics, 16 individuals did not survive after a limb amputation and 19 did. After the discovery of antiseptics, 6 individuals did not survive after a limb amputation and 34 did.

Note that, while you can switch the order of the variables in the `table` function above and still have a perfectly useful table (with the rows and columns switched), we have chosen this particular ordering because it will be helpful when we create our bar chart in the next deliverable.

Deliverable 8: Draw a side-by-side bar chart to show the number of patients who survived and did not survive a limb amputation before and after the discovery of antiseptics.

The `barplot` function can also make side-by-side and stacked bar charts when you want to compare the frequency of one variable grouped by another. This is done plotting the results of a two-by-two table such as the one we created in Deliverable 7. The first variable in the `table` function will be the one graphed on the y-axis, and the second will be used as categories to create the stacked or

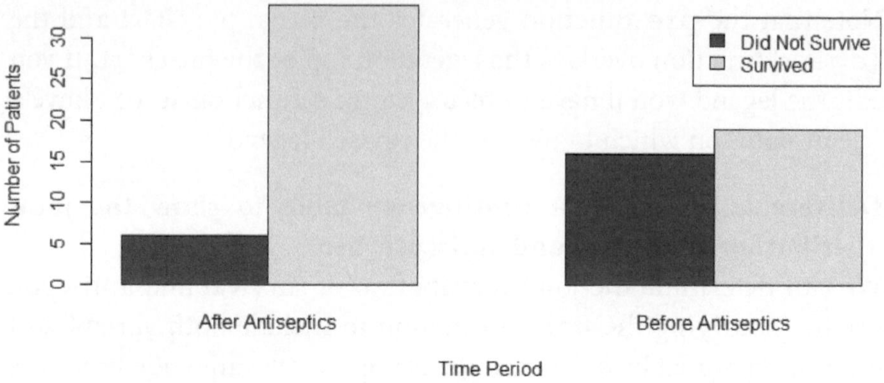

FIGURE 2.4 Patient survival by antiseptic use.

side-by-side bars. Because we want to compare survival grouped by whether the amputation was performed before or after the discovery of antiseptics, the table in the `barplot` function should read `table(Lister$Outcome.char, Lister$Antiseptic.char)` (which is what we wrote in Deliverable 7).

The following R code will generate the side-by-side bar chart of patient outcome by time period shown in Figure 2.4. The argument `beside = TRUE` indicates that we want a side-by-side rather than stacked bar chart, and the argument `legend.text = TRUE` will automatically add a legend to the bar chart.

```
barplot(
   table(Lister$Outcome.char, Lister$Antiseptic.char),
   main="Patient Survival by Antiseptic Use",
   beside = TRUE,
   legend.text = TRUE,
   xlab = "Time Period",
   ylab = "Number of Patients"
      )
```

Deliverable 9: Summarize your findings about the relationship between antiseptic use and survival after an amputation.
Before the discovery of antiseptics, about half of the patients who had a limb amputation survived and half did not survive. However,

after the discovery of antiseptics, many more individuals survived after an amputation than did not survive.

STUDENT PROJECT: DETERMINE THE IMPACT TICKET CLASS HAD ON PASSENGER SURVIVAL ON THE *TITANIC*

In the early hours of April 15, 1912, the unsinkable ship *RMS Titanic* sank when it struck an iceberg, killing more than half of the passengers and crew aboard. The `Titanic.xlsx` dataset contains demographic information for 889 of those passengers as well as a record of whether or not each passenger survived[2]. Our goal is to describe the distributions of passenger ticket class and passenger survival.

Project deliverables:

1. Use the data dictionary to identify each variable in the dataset as categorical or quantitative. If the variable is categorical, further identify it as ordinal, nominal, or an identifier variable.
2. Using the `ifelse` function covered in Chapter 1, create a new variable called `Survived.char` that takes on the value `Did Not Survive` when `Survived == 0` and `Survived` when `Survived == 1`.
3. Calculate the frequency and relative frequency of survival for *Titanic* passengers.
4. Calculate the frequency and relative frequency of passengers in each ticket class.
5. Create a bar chart to display the number of passengers who survived and did not survive the *Titanic* sinking.
6. Create a pie chart to display the percent of passengers who traveled under each ticket class.
7. Create a contingency table to show the joint distribution of survival and passenger ticket class.

8. Draw a side-by-side bar chart to show the number of passengers in each ticket class who survived and did not survive. You can use the argument `args.legend = list(x="topleft")` to move the legend to the top left of the graph.

9. Summarize your findings about survival on the *Titanic* and how a passenger's ticket class might have affected his or her survival.

REFERENCES

1. Lister. (1870). "Effects of the Antiseptic System of Treatment Upon the Salubrity of a Surgical Hospital." *The Lancet*, Vol. 1, no. 4–6, pp. 40–42.
2. Stanford University. (2016). A *Titanic* Probability. *Titanic* Training Dataset. https://web.stanford.edu/class/archive/cs/cs109/cs109.1166/problem12.html

CHAPTER 3

Describing Quantitative Data

INTRODUCTION

Most introductory statistics books and classes focus primarily on analyzing quantitative data, and this book will be no exception. Quantitative data is data that is collected using some sort of measurement. The length of a human pregnancy, the number of lottery winners in a given week, and the frequency and amplitude of a bird's song are all examples of quantitative data.

Quantitative variables can be divided into two broad types: discrete and continuous. Discrete variables have values that can be counted individually in a finite amount of time. The number of people sitting on a bus is a discrete variable. There are either 19 people or 20 people on the bus – there could never be 19.35674689 bus riders.

Continuous variables can take on any value and can be measured ever and ever more precisely. A man may say he's 6 feet tall, but stand a very precise measuring stick next to him and you'll discover that he's really 5 feet 11.2349863 inches tall. If you stood an even more precise measuring stick next to him, you could carry out the measurement to an even finer degree.

STATISTICAL DISTRIBUTIONS

As with categorical variables, the first goal of descriptive statistics for quantitative data is to understand the *distribution* or possible values of the variable and how frequently they occur. For quantitative data, this can be accomplished in two ways: graphically with histograms and boxplots and by using numerical summaries. It's useful to proceed in that order because what you see in your histogram or boxplot will inform the best numerical summary to compute.

Another important feature of a quantitative variable is the presence (or absence) of outliers. Students sometimes think of "outlier" as a synonym for "error"; however, an outlier is just a particularly high or low value, often defined as being larger than 1.5 times the interquartile range (IQR) above the third quartile or smaller than 1.5 times the IQR below the first quartile.

The presence of one or more outliers can signify a number of things. First, there are certain statistical distributions where outliers are expected. A variable that has a high frequency of small values and a few high outliers may have what is known as a Poisson distribution (a more thorough discussion of the Poisson distribution is beyond the scope of this book). The outliers are not unexpected at all but are a normal feature of the data.

Second, the presence of an outlier could represent unusual conditions. For example, the data point for a CEO might show up as an outlier in the distribution of salaries at a company. The CEO salary is not an error but might indicate that the CEO is compensated based on a different set of "rules" that the rest of the staff.

Finally, an outlier might be caused by an error, and these outliers should be fixed or eliminated from analyses. The best way to detect errors in a dataset is to learn more about what each variable is measuring through reading credible sources and talking with subject matter experts.

Say you are analyzing the results of a blood test measuring HDL cholesterol in humans. You should know (not necessarily off the top of your head, but you should be able to easily reference) the "normal" range of HDL cholesterol and the values that would be physiologically impossible. A lab result that is physiologically impossible warrants a second look (could there be a data entry error?) and consultation with a subject matter expert to determine how to proceed.

The distribution of a variable also informs the appropriate summary measures and statistical tests for that variable. Data that has the classic "bell curve" shape (or close to it) is called Normally distributed and is best summarized using the mean for the center and the standard deviation for the spread. Data that has pretty much any other shape (including data with outliers) is best summarized using the median for the center and the range or IQR for the spread. For clarity, we will use "normal" as a synonym for "usual" or "typical" and "Normal" when referring specifically to the Normal Distribution.

GUIDED PROJECT 1: DETERMINE THE CASUALTY RATE FOR UNION ARMY SOLDIERS IN COMPANIES FROM OHIO DURING THE US CIVIL WAR

Approximately 618,000 Union and Confederate soldiers died in battle and from starvation and disease during the single bloodiest conflict in United States history: The Civil War. To put this in perspective, "only" about 400,000 US soldiers perished during the second-most-deadly conflict, World War II. Despite the terrible conditions soldiers faced in the field, remarkably good records about these men, their backgrounds, and their fates have been preserved.

The `Ohio.xlsx` dataset contains records from 45 companies from Ohio during the United States Civil War including the number of soldiers in the company, the year the company was formed, details about the demographic make-up of each company as well as the overall mortality and mortality due to injury and illness[1]. We wish to describe the mortality rates for these companies from Ohio during the US Civil War.

Project deliverables:

1. Use the data dictionary to identify each variable in the dataset as categorical or quantitative. If the variable is categorical, further identify it as ordinal, nominal, or an identifier variable. If the variable is quantitative, identify it as discrete or continuous.

2. Create and describe a histogram showing the distribution of total mortality for the 45 Ohio companies.

3. Based on the shape of the distribution of total mortality, determine and calculate the appropriate measures of center and spread for the data. How many soldiers did a company "typically" lose?

4. Create and describe a histogram showing the distribution of the number of soldiers in each of the 45 Ohio companies.

5. Using the definition of an outlier as any point more than 1.5 times the IQR above Q3 or below Q1, explore any outliers present.

6. Calculate and describe the percent mortality in the 45 Ohio companies.

7. Draw conclusions about the casualty rate for Union Army soldiers in companies from Ohio during the US Civil War.

Deliverable 1: Use the data dictionary to identify each variable in the dataset as categorical or quantitative. If the variable is categorical, further identify it as ordinal, nominal, or an identifier variable. If the variable is quantitative, identify it as discrete or continuous.

Using the methods we learned in Chapter 1, import the `Ohio.xlsx` dataset. Also open an R Script and the "Ohio Data Dictionary" file.

The first variable in the dataset is `Company`. `Company` is a categorical variable that provides the name of each unique company; therefore, it is an identifier variable. The second variable is `No_soldiers`, the number of soldiers in the company. This variable is quantitative because it measures the number of soldiers in the company, and is discrete because "number of soldiers" is something that can be counted in a finite amount of time. The third variable is `Enlist_yr`, which is a categorical variable that identifies each company as having been formed in 1861, 1862, 1863, or 1864. Because it only makes sense to think of those years in that order, `Enlist_yr` is an ordinal variable. The fourth, fifth, and sixth variables are `Death_total`, `Death_illness`, and `Death_injury`. Like `No_soldiers`, these are discrete, quantitative variables that count the total number of deaths, the number of deaths from illness, and the number of deaths from injury, respectively, in each company. The fifth variable is `Pct_farmers`, which is a continuous, quantitative variable that measures the percent of soldiers in each company who were farmers. Percent is a continuous measure because it can take on values like 36.52% and can't be counted like number of deaths and number of soldiers can. Similarly, the last variable, `Pct_foreign`, is also a continuous, quantitative variable.

Deliverable 2: Create and describe a histogram showing the distribution of total mortality for each of the 45 Ohio companies.

Histograms are created using the `hist` function. In an R Script, type and run the statement below to produce Figure 3.1.

```
hist(
     x = Ohio$Death_total,
     main = "Total Mortality: Ohio Companies, US Civil
             War",
     xlab = "Total Number of Deaths",
     ylim = c(0, 20)
    )
```

The first argument, `x = Ohio$Death_total`, is the variable we wish to use to create the histogram. As with the bar charts and pie charts we created in Chapter 2, `main =` allows you to specify the main title of the plot, and `xlab =` and `ylab =` specify the x-axis and y-axis labels. You can also change the range of the y-axis using `ylim = c(smallest value, largest value)`. The range of the x-axis can be modified similarly using `xlim =` .

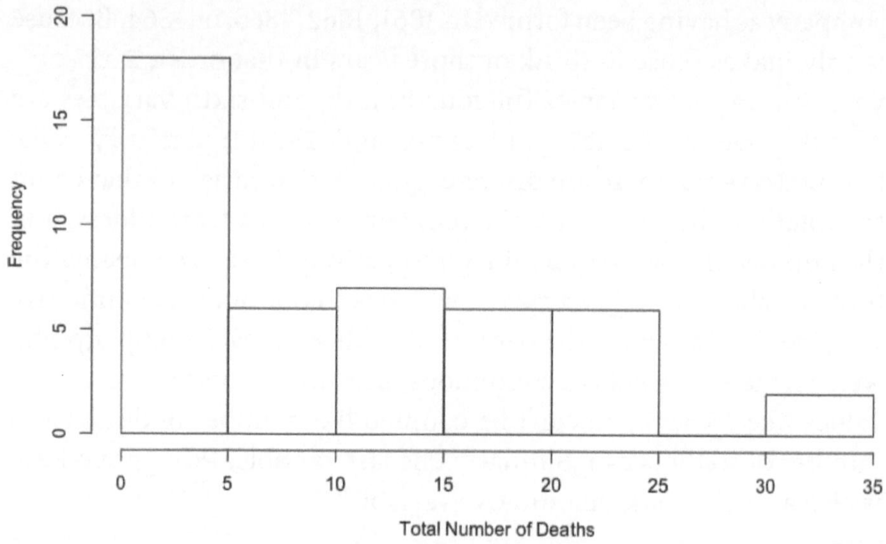

FIGURE 3.1 Total mortality: Ohio companies, US Civil War.

Deliverable 3: Based on the shape of the distribution of total mortality, determine and calculate the appropriate measures of center and spread for the data. How many soldiers did a company "typically" lose?

The distribution of total mortality is skewed to the right with most companies sustaining between one and five casualties and just a few sustaining between 30 and 35. Because the distribution of the data is clearly not Normal, the median will be a better descriptor of the "typical" number of casualties and the IQR will be a better descriptor of how variable the number of casualties was among the companies.

Type and run the command `fivenum(Ohio$Death_total)` in your R Script to calculate the five-number summary for `Death_total` in the order min, Q1, median, Q3, max.

```
> fivenum(Ohio$Death_total)
[1]  0  4  7 17 34
```

From this, we see that these Ohio companies "typically" lost 7 men, with some losing as few as 0 and as many as 34. The middle 50% of companies lost between 4 and 17 soldiers.

In comparison, the mean number of deaths (calculated using the `mean` function) was quite a bit higher than the median.

```
> mean(Ohio$Death_total)
[1] 10.75556
```

The skewed nature of the data makes the mean a less accurate measure of center than the median.

While completing this deliverable goes a long way toward answering the research question posed in the project description, it leaves out an important piece of context: How big were these companies to begin with? Every individual death is a tragedy, but losing 34 soldiers out of 10,000 is much different from a military perspective than losing 34 out of 70.

Deliverable 4: Create and describe a histogram showing the distribution of the number of soldiers in each of the 45 Ohio companies.

To do this, we will use the `hist` function again. Using methods we've seen before, we give this histogram a title, modify the x- and y-axes and label the x-axis. Type and run the following command in your R Script to produce Figure 3.2:

```
hist(
     x = Ohio$No_soldiers,
     main = "Number of Soldiers: Ohio Companies, US
             Civil War",
     ylim = c(0,20),
     xlim = c(0,200),
     xlab = "Number of Soldiers in each Company"
     )
```

From the histogram, we see that most of the companies had close to 100 soldiers, in some cases more than 150. In at least one case, a company had very few soldiers, possibly less than 10.

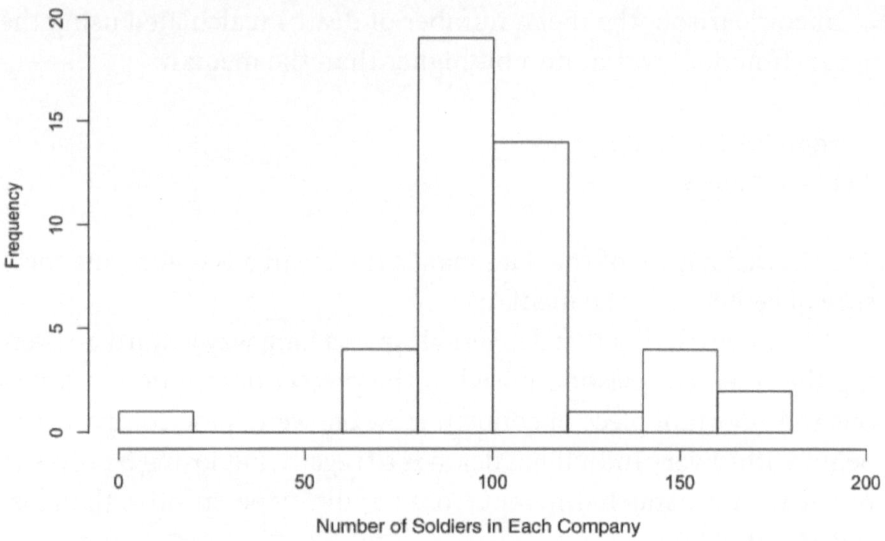

FIGURE 3.2 Number of soldiers: Ohio companies, US Civil War.

Deliverable 5: Using the definition of an outlier as any point more than 1.5 times the IQR above Q3 or below Q1, explore any outliers present.

We can use the results of the five number summary to identify potential outliers in the data. Running the relevant code produces the following output:

```
> fivenum(Ohio$No_soldiers)
[1]  10  87 100 111 180
```

The IQR is calculated as `111 - 87 = 24`. Any observation smaller than `51 (Q1 - 1.5 * 24)` or larger than `147 (Q3 + 1.5 * 24)` is considered an outlier. Both the minimum and maximum values (and possibly other points) are outliers. Again, "outliers" aren't necessarily errors. They are just points that are just particularly large or small compared to the bulk of the data.

Creating a box plot in addition to a histogram allows us to view the distribution of `No_soldiers` in a different way. Of particular interest for this deliverable is that boxplots identify observations that are outliers as little circles.

The boxplot function displays the distribution of a variable using a boxplot. We see the same statements used to specify the title of the plot and the label for the x-axis. One notable addition is the use of horizontal = TRUE which orients the boxplot horizontally rather than vertically. This is simply an aesthetic choice. Running the following statements generates Figure 3.3.

```
boxplot(
     x = Ohio$No_soldiers,
     main = "Number of Soldiers: Ohio Companies, US
             Civil War",
     horizontal = TRUE,
     xlab = "Number of Soldiers in Each Company"
        )
```

We see from the circles appearing on the boxplot that there is at least one observation on the small side of the distribution and several observations on the large side that are considered outliers.

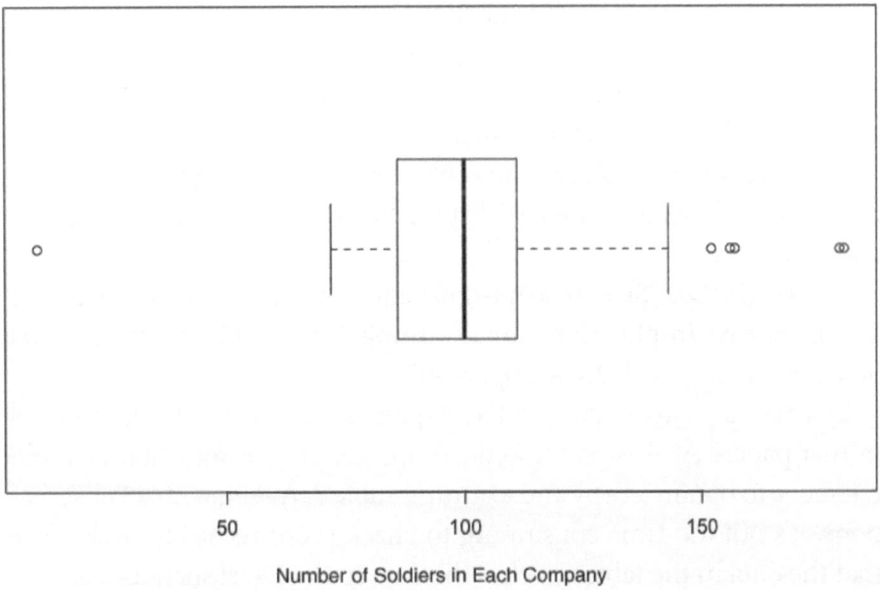

FIGURE 3.3 Number of soldiers: Ohio companies, US Civil War.

At this point, we should be curious about the very large and very small observations. But what to do? There are two possibilities for what is going on here: The first is that these observations are unusual but correct. There were very large and very small companies in Ohio during this period of the US Civil War. The second possibility is that there is an error in the data or data analysis. We'll explore the second case – that the unusual observations are the result of errors – first and then, if we can't find any errors, we'll do what we can to confirm that the outliers are correct. When investigating errors, it's useful to start at the point you noticed the potential error and then work backwards toward the original source of the data to see if you can identify where an error might have been introduced.

The first thing to consider is that we made an error with our plots or calculations. We can run `View(Ohio)`, and, because there are only 45 observations, we can scroll through the spreadsheet view of the data and see that there is one company with 180 soldiers and one with only 10 soldiers. We have not made an error plotting or calculating the summary measures.

The next possibility is that there was an error reading in the data from Excel into RStudio. To check this, we can open the `Ohio.xlsx` file and look at the data there. Again, we see Company 78D has 180 soldiers and Company 195B has 10 soldiers. So there is not an error reading in the dataset from Excel.

Now we must consider a possible error converting the data from its previous form into the Excel spreadsheet. The project description cites the paper:

C. Lee (1999). "Selective Assignment of Military Positions in the Union Army: Implications for the Impact of the Civil War", Social Science History, Vol. 23, #1, pp. 67-97.

Does the data in our current Excel spreadsheet match what is reported in that paper? At this point, I encourage you to use your library's web reference to obtain a copy and examine Table 3. Again, with 45 observations, it's not too time consuming to check every record to make sure that the data in the table matches the data in the Excel spreadsheet.

We see in Table 3 of the paper that nearly all of the records for "number of soldiers" match, including Company 78D with 180 soldiers.

However, if we look at the second page of Table 3, we see that Company 195B has 100 soldiers, not 10 soldiers. We have found an error, and we can confidently edit the `Ohio.xlsx` dataset to change the number of soldiers in Company 195B to 100. Remove the old, incorrect version of the Ohio dataset from RStudio by typing and running `rm(Ohio)` in your R Script. Then import the corrected version.

What about the larger companies? The fact that they appear in a table of a peer-reviewed publication should give us confidence that they are correct. Dr. Lee does not comment in this paper about the differing size of the companies. He does, however, mention statistical differences in mortality between companies that were formed earlier and later in the war. A quick look at the data indicates that the larger companies were formed earlier rather than later in the war. This seems to be a reasonable explanation for why the companies were larger. They were simply in existence longer and had more opportunity for men to be assigned to them. At this point, we can conclude with confidence that the large observations are unusual but correct.

Using the corrected version of the dataset, we can examine the distribution of the size of each company using a histogram (Figure 3.4), boxplot (Figure 3.5), and calculated summary measures.

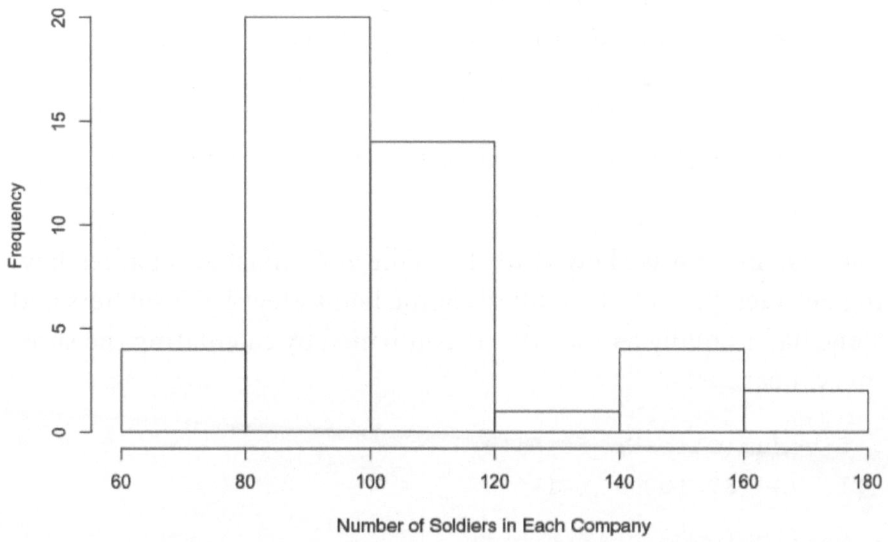

FIGURE 3.4 Number of soldiers: Ohio companies, US Civil War.

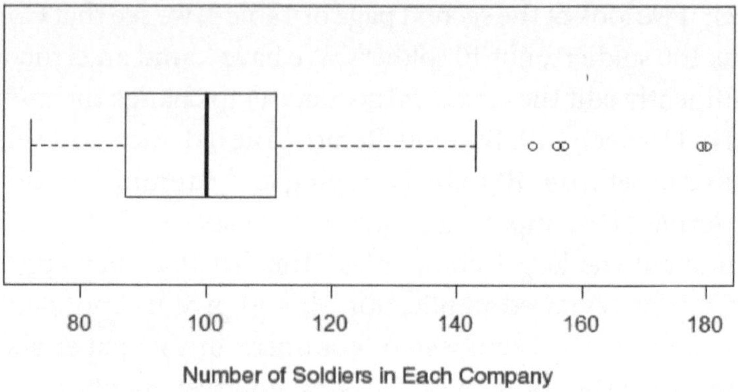

FIGURE 3.5 Number of soldiers: Ohio companies, US Civil War.

```
hist(
     x = Ohio$No_soldiers,
     main = "Number of Soldiers: Ohio Companies, US
             Civil War",
     ylim = c(0,20),
     xlab = "Number of Soldiers in Each Company"
     )

boxplot(
     x = Ohio$No_soldiers,
     main = "Number of Soldiers: Ohio Companies,
             US Civil War",
     horizontal = TRUE,
     xlab = "Number of Soldiers in Each Company"
        )
```

The distribution is skewed to the right with most companies having between 87 and 111 soldiers. Some had as few as 72 soldiers and some had as many as 180. This is confirmed by calculating the summary measures.

```
> fivenum(Ohio$No_soldiers)
[1]  72  87 100 111 180

> mean(Ohio$No_soldiers)
[1] 105
```

Again, we note that the median is a better measure of center than the mean because the data is skewed.

Deliverable 6: Calculate and describe the percent mortality in the 45 Ohio companies.

Although our original question wasn't phrased this way, we are beginning to see what we are really trying to answer isn't, "How many casualties were there in these Ohio companies?". Because there was quite a bit of variation in company size, we need to answer the question "What percent of the soldiers in each company died during the war?". This is often true in statistical analyses. The relative frequency (or percent) of an occurrence often tells us more than the raw frequency (or count).

To gain insight into the relative frequency of deaths due to any cause, we can create a new variable by dividing the total number of deaths by the number of soldiers in each company and multiplying by 100%.

```
Ohio$Pct.dead <-(Ohio$Death_total / Ohio$No_soldiers)
                    * 100
```

Once we've added the variable to the Ohio dataset, it should be added to the data dictionary as well.

We can now create a histogram and calculate summary statistics for the percent of soldiers who died from all causes. Type and run the following command in an R Script to create Figure 3.6:

```
hist(
     x = Ohio$Pct.dead,
     main = "Percent casualties: Ohio Companies, US
             Civil War",
     xlab = "Percent Dead"
     )
```

From the histogram, we see the distribution of the data is skewed right. Most companies had less than a 10% mortality rate but one company had at least a 30% mortality rate.

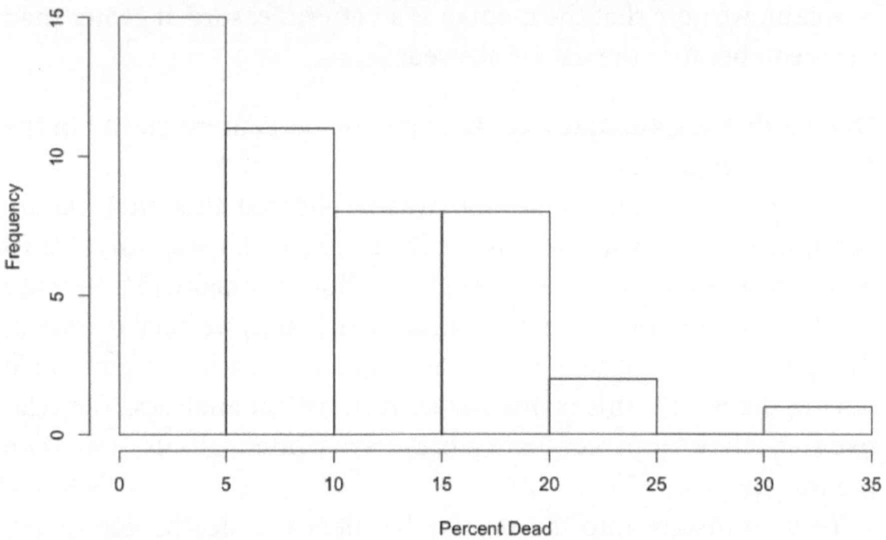

FIGURE 3.6 Percent casualties: Ohio companies, US Civil War.

We can use the `fivenum` function to further quantify the death rate.

```
> fivenum(Ohio$Pct.dead)
[1]  0.000000  4.494382  6.896552 14.545455 30.088496
```

Deliverable 7: Draw conclusions about the casualty rate for union army soldiers in companies from Ohio during the US Civil War. A company from Ohio typically had a casualty rate of about 7%, and 25% of companies lost fewer than 5% of their men. Only 25% of companies had a casualty rate of more than 14%, but one company lost 30% of its men. Note that the skewed nature of the data makes the median (7%) a better measure of the "typical" casualty rate than the mean (more than 9%).

GUIDED PROJECT 2: DETERMINE THE TYPICAL HEALTHY ADULT HUMAN BODY TEMPERATURE

Everyone knows that 98.6°F (37.0°C) is the normal human body temperature. But is that actually correct, and – come to think of it – how does *everyone* know that in the first place?

A German physician named Carl Reinhold August Wunderlich is generally credited with originating this idea, which was based on – reportedly – more than one million axillary temperature readings taken from 25,000 subjects and was published in his 1868 book *Das Verhalten der Eigenwärme in Krankheiten* (which translates to *The Behavior of the Self-Warmth in Diseases*). But was he correct? History tells that his thermometer was a foot long and took 20 minutes to determine a subject's temperature. For a measure that is used so often to determine general health, it would be a good idea to use modern instruments to confirm or refute his results.

In 1992, three physicians from the University of Maryland School of Medicine set out to do just that, measuring body temperatures for 223 healthy men and women aged 18–40 one to four times a day for 3 consecutive days using an electronic digital thermometer. The mean body temperature was computed for each individual, and this summary measure is recorded in the `Bodytemp.xlsx` dataset[2]. We wish to describe the "typical" body temperature for the participants in his study.

Project deliverables:

1. Import the `Bodytemp.xlsx` dataset into RStudio. Use the data dictionary to identify each variable in the dataset as categorical or quantitative. If the variable is categorical, further identify it as ordinal, nominal, or an identifier variable. If the variable is quantitative, identify it as discrete or continuous.

2. Create and describe a histogram showing the distribution of body temperature for the participants in the study.

3. Based on the shape of the distribution of body temperature, determine and calculate the appropriate measures of center and spread for the data.

4. Draw a boxplot of the data and explore any unusual values or outliers present. Determine if these values should be included in or excluded from analyses.

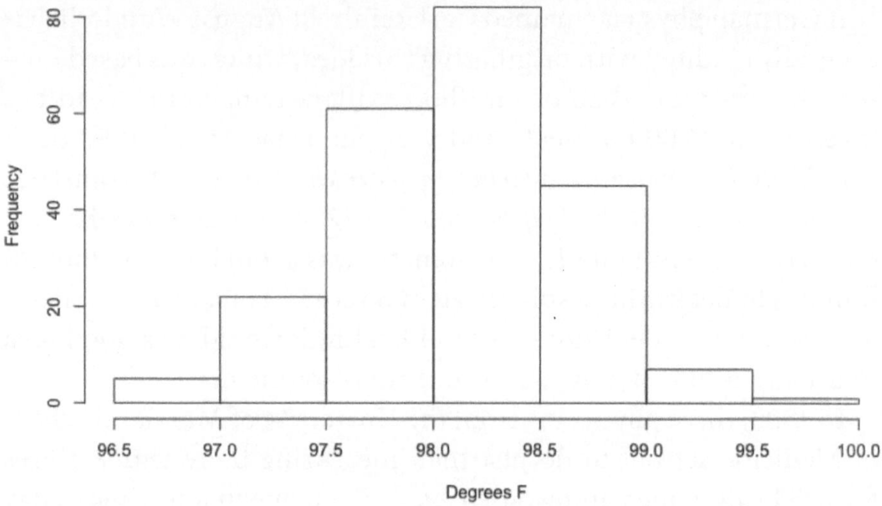

FIGURE 3.7 Body temperature.

Deliverable 1: Import the `Bodytemp.xlsx` dataset into RStudio. Use the data dictionary to identify each variable in the dataset as categorical or quantitative. If the variable is categorical, further identify it as ordinal, nominal, or an identifier variable. If the variable is quantitative, identify it as discrete or continuous.

Import the `Bodytemp.xlsx` data and open an R Script and the file "Bodytemp Data Dictionary". There are only two variables in the dataset. The first is `ID`, which is a categorical identifier variable used to identify each participant in the study. The second is `Body_temp`. This is a continuous, quantitative variable measured in °F.

Deliverable 2: Create and describe a histogram showing the distribution of body temperature for the participants in the study.

Type and run the following command in your R Script to create Figure 3.7:

```
hist(
     x = Bodytemp$Body_temp,
     main = "Body Temperature",
     xlab = "Degrees F"
     )
```

Deliverable 3: Based on the shape of the distribution of body temperature, determine and calculate the appropriate measures of center and spread for the data.

Because the histogram of body temperature is unimodal and symmetric (Normal), the mean and standard deviation are the appropriate measures of center and spread. Running the `mean` and `sd` functions shown below produces the following output:

```
> mean(Bodytemp$Body_temp)
[1] 98.15919

> sd(Bodytemp$Body_temp)
[1] 0.5237612
```

In this sample, the mean body temperature was 98.16°F with a standard deviation of 0.52°F.

Deliverable 4: Draw a boxplot of the data and explore any unusual values or outliers present. Determine if these values should be included in or excluded from analyses.

Type and run the following in your R Script to generate Figure 3.8:

```
boxplot(
     x = Bodytemp$Body_temp,
     main = "Body Temperature",
     xlab = "Degrees F",
     horizontal = TRUE
   )
```

From the boxplot we see that there are two outliers, one on the high end and one on the low end. Here is where literature review or consultation with a subject matter expert is helpful. Both of these values are physiologically possible (unlike, say, a human body temperature of 30°F or 200°F) so they are unusual, but we don't have any reason to suspect they are errors.

While these temperature readings were taken in healthy adults, it is also possible to be ill without knowing it, and the two "healthy"

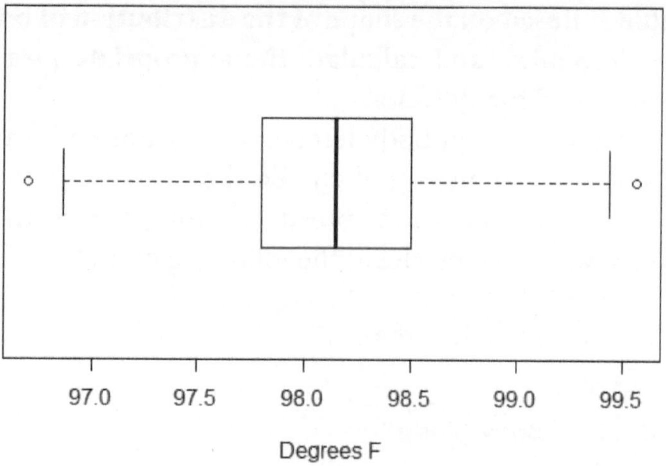

FIGURE 3.8 Body temperature.

outliers in the study may have had an unknown condition that reduced or elevated their body temperatures. However, without more information, we have no reason to treat these data points as errors.

STUDENT PROJECT 1: DETERMINE THE TYPICAL WEIGHT OF AN ADULT HUMAN BRAIN

The `Brainhead.xlsx` dataset provides information on 237 individuals who were subject to postmortem examination at the Middlesex Hospital in London around the turn of the 20th century[3]. Study authors used cadavers to see if a relationship between brain weight and other, more easily measured, physiological characterizes such as age, sex, and head size could be determined. The end goal was to develop a way to estimate a person's brain size while they were still alive (as the living aren't keen on having their brains taken out and weighed). We wish to use this data to determine the weight of a "typical" human brain.

Project deliverables:

1. Import the `Brainhead.xlsx` dataset into RStudio. Use the data dictionary to identify each variable in the dataset as categorical or quantitative. If the variable is categorical, further identify it as ordinal, nominal, or an identifier variable. If the variable is quantitative, identify it as discrete or continuous.

2. Draw a histogram and a boxplot displaying the distribution of brain weight.
3. Based on the shape of the distribution of brain weight, determine and calculate the appropriate measures of the center and spread of the data.
4. Draw conclusions about the weight of a typical adult human brain.

STUDENT PROJECT 2: DESCRIBE THE DISTRIBUTION OF HORSE KICK DEATHS IN THE PRUSSIAN ARMY FROM 1875 TO 1894

From its inception in the 1600s until it was replaced by the Imperial Germany Army in the late 19th century, the Prussian Army was one of the most victorious fighting forces on the planet. However, the strength of any army depends on its smallest components – in this case, its infantrymen and its horses. Because horses kick when they are feel threatened or hurt or to show dominance, a death resulting from a horse kick is not only a loss of a soldier but also possibly an indication that the horse has been mistreated.

Russian economist Ladislaus Bortkiewicz first published data on deaths from horse kicks in the Prussian Army in his 1898 statistics book entitled *Das Gesetz der kleinen Zahlen* (which translates to *The Law of Small Numbers*)[4]. The `Horsekicks.xlsx` dataset contains the records from his analysis of the number of deaths from horse kicks that occurred in 14 different Prussian Army corps from 1875 to 1894. As a point of interest, using methods that are beyond the scope of this chapter, Bortkiewicz determined that deaths from horse kicks in the Prussian army were random.

Project deliverables:

1. Import the `Horsekicks.xlsx` dataset into RStudio. Use the data dictionary to identify each variable as categorical or quantitative. Identify each categorical variable as ordinal or nominal. Identify each quantitative variable as continuous or discrete.

2. Create a histogram and a boxplot to display the distribution of the number of deaths from horse kicks in the 14 Prussian Army corps from 1875 to 1894.

3. Based on the shape of the distribution of horse kick casualties, determine and calculate the appropriate measures of the center and spread of the data.

4. Draw conclusions about the frequency of horse kick deaths in the Prussian Army during this period.

REFERENCES

1. Lee, C. (1999). "Selective Assignment of Military Positions in the Union Army: Implications for the Impact of the Civil War," *Social Science History*, Vol. 23, no. 1, pp. 67–97.
2. Mackowiak, P. A., Wasserman, S. S., and Levine, M. M. (1992). "A Critical Appraisal of 98.6 Degrees F, the Upper Limit of the Normal Body Temperature, and Other Legacies of Carl Reinhold August Wunderlich," *Journal of the American Medical Association*, Vol. 268, pp. 1578–1580.
3. Gladstone, R. J. (1905). "A Study of the Relations of the Brain to the Size of the Head," *Biometrika*, Vol. 4, pp. 105–123.
4. von Bortkiewicz, L. (1898). *Das Gesetz der kleinen Zahlen*. Leipzig: B.G. Teubner.

CHAPTER 4

The Normal Distribution

INTRODUCTION

Of all the statistical distributions, perhaps none is more iconic than the Normal distribution. Even people who have never taken a statistics class may recognize the famous "bell-shaped" curve and have a pretty good intuitive idea what it means. When a variable has a Normal distribution, most observations are right around the mean and a few (in equal numbers) are much higher or much lower.

One of the goals of statistics is to identify and investigate both the "usual" and "unusual" in the world. In order to know if someone's blood pressure is too high, we must first have a notion of what "normal" blood pressure is. In order to know if a part in a jet engine is unsafe, we first have to understand the "typical" failure rate of that part.

When a variable does fit a Normal model, it gives us a powerful tool to describe what is usual and unusual for that variable. The mean (because it is equal to both the median and mode) is a good descriptor of a "typical" or "usual" value for the variable. The standard deviation is a ruler we can use to measure how "near" or "far" an observation is from the mean. This provides a standard method for determining when we've moved from the realm of the expected to the realm of the unexpected.

Later on, when we study statistical inference, the mathematical properties of the Normal model will allow us to assign a value of how likely or unlikely we believe an event to be based on our preconceived notion of "usual". For now, we'll look at describing and displaying data that has a Normal distribution and begin to develop an intuitive idea of how it can be used to discriminate between the expected and unexpected. For clarity, we will use "normal" as a synonym for "usual" or "typical" and "Normal" when referring specifically to the Normal Distribution.

GUIDED PROJECT: DETERMINE THE SPEED OF LIGHT IN AIR

The speed of light is an important physical constant in physics and astronomy. Although it was determined as early as the 1600s that light traveled with a speed (as opposed to instantaneously), the exact speed had still not been measured in the late 19th century.

Building on the work of others, Albert A. Michaelson conducted a series of experiments using a device with a complicated system of lenses and revolving mirrors, and in June and early July 1879, he obtained 100 measurements of the speed of light. His results were first published in a report to the United States Secretary of the Navy[1].

We wish to use Michaelson's original data, found in the `Michaelson.xlsx` dataset, to estimate the speed of light in air and compare it to the true speed of light in air, which was later determined to be 299,700 km/s.

Project deliverables:

1. Import the `Michaelson.xlsx` dataset into RStudio. Use the data dictionary to identify each variable in the dataset as categorical or quantitative. If the variable is categorical, further identify it as ordinal, nominal, or an identifier variable. If the variable is quantitative, identify it as discrete or continuous.

2. Draw a histogram displaying the distribution of measurements of the speed of light in air produced by Michaelson's apparatus. Does a Normal model seem to fit this data?

3. Draw a boxplot displaying the distribution of the measurements of the speed of light in air produced by Michaelson's apparatus.

4. Describe the distribution of the measurement of the speed of light produced during Michaelson's experiments.
5. Use the 68-95-99.7% rule to determine the range of the middle 95% of measures of the speed of light we would expect to be generated by Michaelson's apparatus.
6. Calculate a z-score for the true speed of light in air (299,700 km/s) using the mean and standard deviation generated by Michaelson's experiments.
7. What proportion of the time would we expect Michaelson to have measured a value of the speed of light that is greater than the true value (299,700 km/s)?
8. Draw conclusions about Michaelson's experimental results and what we now know to be the true speed of light in air.

Deliverable 1: Import the `Michaelson.xlsx` dataset into RStudio. Use the data dictionary to identify each variable in the dataset as categorical or quantitative. If the variable is categorical, further identify it as ordinal, nominal, or an identifier variable. If the variable is quantitative, identify it as discrete or continuous.

Using the methods we learned in Chapter 1, import the `Michaelson.xlsx` dataset. Also open the file "Michaelson Data Dictionary" and an R Script. The first variable is `ID`, which is a categorical identifier variable that identifies each of Michaelson's observations uniquely. The second variable is `speedoflight`. This is a continuous, quantitative measure of speed of light through air measured in kilometers per second.

Deliverable 2: Draw a histogram displaying the distribution of Michaelson's measurements of the speed of light in air. Does a Normal model seem to fit this data?

Type and run the following histogram function in an R Script to create Figure 4.1:

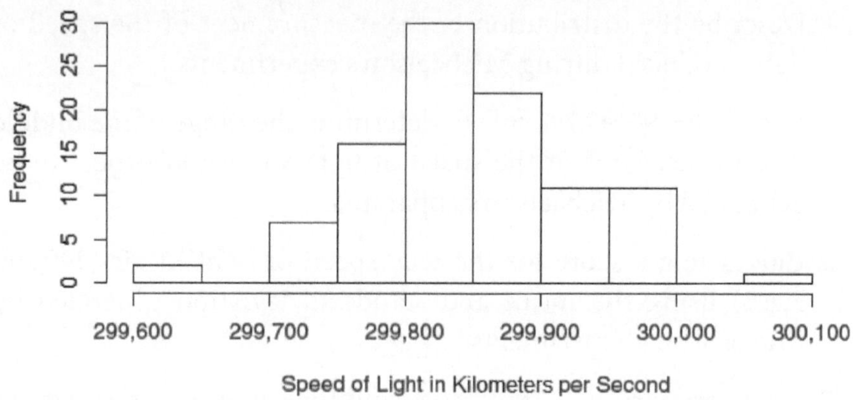

FIGURE 4.1 Measurements of the speed of light in air.

```
hist(
    x = Michaelson$speedoflight,
    main = "Measurements of the Speed of Light in Air",
    xlab = "Speed of Light in Kilometers per Second",
    ylim = c(0,30)
    )
```

The distribution is unimodal and symmetric with two possible outliers – one on the left side of the distribution and one on the right. A Normal model seems to fit this data.

Deliverable 3: Draw a boxplot displaying the distribution of measurements of the speed of light in air produced by Michaelson's apparatus.

Type and run the following boxplot function in an R Script to create Figure 4.2:

```
boxplot(
    x = Michaelson$speedoflight,
    main = "Measurements of the Speed of Light in Air",
    xlab = "Speed of Light in Kilometers per Second",
    horizontal = TRUE
      )
```

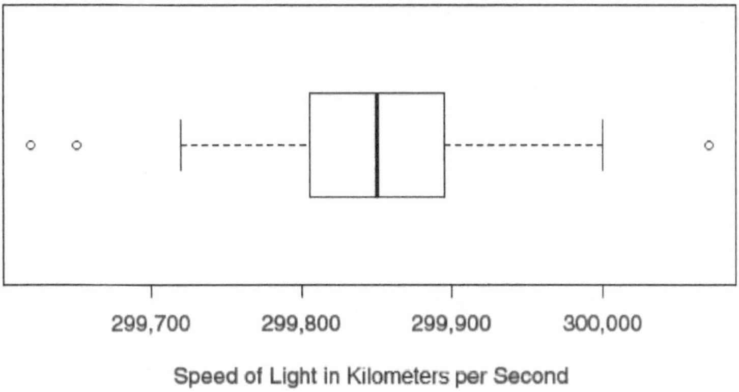

FIGURE 4.2 Measurements of the speed of light in air.

The boxplot confirms our description of the data in Deliverable 2. The distribution appears to be symmetric with two outliers on the left side and one on the right. A review of the original publication confirms these data to be valid experimental results.

Deliverable 4: Describe the distribution of the measurement of the speed of light produced during Michaelson's experiments. Because the distribution of `speedoflight` is unimodal and symmetric, the mean and standard deviation are the best summary measures of the center and spread of the data. Type and run the following functions in an R Script to calculate the mean and SD of the experimental measures of the speed of light, which will produce the results below:

```
> mean.speedoflight <- mean(Michaelson$speedoflight)
> mean.speedoflight
[1] 299852.4

> sd.speedoflight <- sd(Michaelson$speedoflight)
> sd.speedoflight
[1] 79.01055
```

The distribution of measures of the speed of light generated by Michaelson's apparatus appears to be Normal with a mean of 299,852.4 km/s and a standard deviation of 79 km/s.

Deliverable 5: Use the 68-95-99.7% rule to determine the range of the middle 95% of measures of the speed of light we would expect to be generated by Michaelson's apparatus.

The 68-95-99.7% rule tells us that, for a variable with a Normal distribution, the middle 95% of the data will be within two standard deviations of the mean. Type and run the following functions in an R Script to find the range of the mean value of the speed of light plus and minus two standard deviations.

```
> mean.speedoflight - 2 * sd.speedoflight
[1] 299694.4

> mean.speedoflight + 2 * sd.speedoflight
 [1] 300010.4
```

The middle 95% of measurements of the speed of light in air generated by Michaelson's apparatus are expected to be between about 299,694 km/s and 300,010 km/s.

Deliverable 6: Calculate a z-score for the true speed of light in air (299,700 km/s) using the mean and standard deviation generated by Michaelson's experiments.

Type and run the following function in an R Script to find a z-score for the true speed of light in air (299,700 km/s). To calculate the z-score, we subtract the mean of the experimental values from the true speed of light and divide by the standard deviation of the experimental data.

```
> (299700-mean.speedoflight)/sd.speedoflight
[1] -1.928856
```

The z-score for the true speed of light in air is −1.93. The true speed of light is 1.93 standard deviations below the mean generated by Michaelson's experiments.

Deliverable 7: What proportion of the time would we expect Michaelson to have measured a value of the speed of light in air that is greater than the true speed of light in air (299,700 km/s).

The pnorm function allows us to find the proportion of the data in a Normal model that we would expect to be larger or smaller than a particular value. Type and run the following function in an R Script to calculate the proportion of time Michaelson's apparatus would be expected to overestimate the speed of light.

```
> pnorm(
      299700,
      mean = mean.speedoflight,
      sd = sd.speedoflight,
      lower.tail = FALSE
       )
[1] 0.9731257
```

The first number entered in the pnorm function is the value of interest (in this case the true speed of light in air, 299,700 km/s). The argument mean is assigned the sample mean, mean.speedoflight, and the argument sd is assigned the sample standard deviation, sd.speedoflight.

By default, the pnorm function returns the proportion of data we expect to fall to the left of the value of interest. However, in this case we want to find the proportion of the time we expect the apparatus to generate a value of the speed of light *greater* than the true value, so we add the statement lower.tail=FALSE.

Michaelson's apparatus would be expected to generate an estimate of the speed of light that is greater than the true speed of light about 97% of the time.

Deliverable 8: Draw conclusions about Michaelson's experimental results and what we now know to be the true speed of light in air. Michaelson's apparatus generated measurements of the speed of light in air that can be modeled by a Normal distribution with a mean of 299,852.4 km/s and a standard deviation of 79 km/s. In comparison, using modern methods, we are able to measure the speed of light in air as 299,700 km/s. The true speed of light in air is about two standard deviations below the mean estimated using Michaelson's experiments. His apparatus would be expected to overestimate the speed of light about 97% of the time.

STUDENT PROJECT: IS THERE EVIDENCE THAT THE HEALTHY ADULT HUMAN BODY TEMPERATURE IS NOT 98.6°F?

Everyone knows that 98.6°F (37.0°C) is the normal human body temperature. But is that actually correct, and – come to think of it – how does *everyone* know that in the first place?

A German physician named Carl Reinhold August Wunderlich is generally credited with originating this idea, which was based on – reportedly – more than one million axillary temperature readings taken from 25,000 subjects and was published in his 1868 book *Das Verhalten der Eigenwärme in Krankheiten* (which translates to *The Behavior of the Self-Warmth in Diseases*). But was he correct? History tells that his thermometer was a foot long and took 20 minutes to determine a subject's temperature. For a measure that is used so often to determine general health, it would be a good idea to use modern instruments to confirm or refute his results.

In 1992, three physicians from the University of Maryland School of Medicine set out to do just that, measuring body temperatures for 223 healthy men and women aged 18–40 one to four times a day for three consecutive days using an electronic digital thermometer. The mean body temperature was computed for each individual, and this summary measure is given in the `Bodytemp.xlsx` dataset[2].

Project deliverables:

1. Import the `Bodytemp.xlsx` dataset into RStudio. Create and describe a histogram showing the distribution of body temperature for the participants in the study. Is a Normal model appropriate for this data?

2. Draw a boxplot displaying the distribution of body temperature. Does this figure add any additional information to your description of the distribution of the data?

3. Calculate the mean and standard deviation of body temperature for the participants in the study.
4. Using that mean and standard deviation, calculate a z-score for a body temperature of 98.6°F.
5. If healthy human body temperature is described using a Normal model with the mean and standard deviation found in the sample, what proportion of healthy individuals would be expected to have a body temperature less than 98.6°F?
6. Do your results suggest that 98.6°F is not the mean healthy human body temperature?
7. If healthy human body temperature is described using a Normal model with the mean and standard deviation found in the sample, determine the range of the middle 99.7% of values of healthy human body temperature.
8. At what point might a medical professional begin to suspect that an individual is too hot or too cold?

REFERENCES

1. Michelson, A. A. (1882). "Experimental Determination of the Velocity of Light at the United States Naval Academy, Annapolis," *Astronomical Papers*, Vol. 1, pp. 109–145. U.S. Nautical Almanac Office.
2. Mackowiak, P. A., Wasserman, S. S., and Levine, M. M. (1992), "A Critical Appraisal of 98.6 Degrees F, the Upper Limit of the Normal Body Temperature, and Other Legacies of Carl Reinhold August Wunderlich," *Journal of the American Medical Association*, Vol. 268, pp. 1578–1580.

CHAPTER 5

Two-Way Tables

INTRODUCTION

Now that we've learned to display and describe the distribution of a single categorical and a single quantitative variable, the next logical step is to learn to describe the relationship between two variables. Just as the graphs, tables, and numerical summaries used to describe a single categorical variable are different from those used to describe a single quantitative variable, the same is true for relationships between two categorical (the subject of Chapter 5) and two quantitative (the subject of Chapter 6) variables.

CONTINGENCY TABLES

We display the joint distribution of two categorical variables using a contingency table (also called a two-way table or cross-tabulation). This is similar to a frequency table for a single categorical variable except it is modified so that the possible values of one variable are assigned to the columns of the table and the possible values of the other variable are assigned to the rows. Each cell of the table shows the count or percent of observations that have a particular combination of the two variables.

We analyze the relationship between two categorical variables by calculating their joint, marginal, and conditional distributions. The joint distribution shows how often each combination of the two variables occurs and is displayed as the center cells of a contingency table. The marginal distribution shows the frequency of each variable individually, ignoring the other variable, and is displayed as the margin totals in the contingency table (hence the name "marginal distribution").

Finally, the conditional distribution of a variable gives the proportion of observations in each category of one variable for each value of the other variable. This is calculated by dividing the cell count by the row or column total rather than the overall total. Changes in the pattern of the conditional distribution gives information about the relationship between the two variables.

GUIDED PROJECT: ARE FEMALE CHARACTERS IN SLASHER FILMS MORE LIKELY TO BE MURDERED THAN MALE CHARACTERS?

Slasher movies thrill audiences by portraying a lone antagonist (typically male) who attacks innocent victims with extreme violence and without apparent motive. However, this exciting (if gory) subgenre of horror film is criticized by those who view the violence as being used to "punish" female characters who engage in sexual activity during the film. To test this claim, study authors randomly sampled 50 North American slasher films released from 1960 to 2009 and coded the 485 characters appearing in them as being male or female, involved in sexual activity or not, and if they survived the film or not[1]. The data appears in the `Slasher.xlsx` dataset. In this project, we are going to answer a slightly simpler question: Are female characters in slasher films more likely to be murdered than male characters?

Project deliverables:

1. Import the `Slasher.xlsx` dataset into RStudio. Use the data dictionary to identify each variable in the dataset as categorical or quantitative. If the variable is categorical, further identify it as ordinal, nominal, or an identifier variable. If the variable is quantitative, identify it as discrete or continuous.

2. Use the `ifelse` function to create a new variable called `Survival.char` that takes on the value `Survived` when the character survived and `Died` when the character died. Similarly, create another new variable called `Gender.char` that takes on the values `Female` and `Male` for female and male characters, respectively.
3. Describe the frequency and relative frequency of character gender and survival in this set of slasher films.
4. Create a contingency table showing the joint distribution of character survival and gender. Add the table margins to show the marginal distribution of each variable.
5. Modify the contingency table in Deliverable 4 to show the conditional distribution of survival by gender.
6. Create a side-by-side bar chart displaying the conditional distribution of survival by character gender.
7. Draw conclusions about the risk of death for male and female characters in slasher films.

Deliverable 1: Import the `Slasher.xlsx` dataset into RStudio. Use the data dictionary to identify each variable in the dataset as categorical or quantitative. If the variable is categorical, further identify it as ordinal, nominal, or an identifier variable. If the variable is quantitative, identify it as discrete or continuous.
The first variable in the dataset is `ID`, which is a categorical identifier variable that identifies each character uniquely. The second variable is `Gender`, which is a nominal categorical variable identifying each character as being male or female. The third variable is `Activity`, which is a nominal categorical variable indicating if the character was depicted engaging in sexual activity. The fourth variable is `Survival`, which is a nominal categorical variable indicating if the character survived the film or not.

Deliverable 2: Use the `ifelse` function to create a new variable called `Survival.char` that takes on the value `Survived`

when the character survived and `Died` when the character died. Similarly, create another new variable called `Gender.char` that takes on the values `Female` and `Male` for female and male characters, respectively.

Returning to the `ifelse` function we learned in Chapter 1, we can create the `Survival.char` and `Gender.char` variables using the following statements:

```
Slasher$Survival.char <-
     ifelse(
          test = (Slasher$Survival == 0 ),
          yes = 'Died',
          no = 'Survived'
            )

Slasher$Gender.char <-
     ifelse(
          test = (Slasher$Gender == 0 ),
          yes = 'Male',
          no = 'Female'
            )
```

Deliverable 3: Describe the frequency and relative frequency of character gender and survival in this set of slasher films.

We use the `table` and `prop.table` functions (refer to Chapter 2 for more details) to calculate the frequency and relative frequency of character gender and survival in this set of slasher films. Running the relevant code produces the following output:

```
> table(Slasher$Survival.char)
    Died Survived
     400       85

> prop.table(table(Slasher$Survival.char)) * 100
    Died Survived
82.47423 17.52577
```

Only about 18% of the characters in the slasher films survived to the end of the movie. About 82% did not.

Running the relevant R code to calculate the frequency and relative frequency of character gender in this set of slasher films produces the following results:

```
> table(Slasher$Gender.char)
Female   Male
   222    263

> prop.table(table(Slasher$Gender.char)) * 100
 Female         Male
45.7732      54.2268
```

Movie characters were relatively equally split between the genders. About 46% were female and 54% were male.

Deliverable 4: Create a contingency table showing the joint distribution of character survival and gender. Add in the table margins to show the marginal distribution of each variable as well. We create the contingency table to display the relationship between character gender and survival by including both variables in the `table` function separated by a comma. The first variable entered is displayed as the row variable and the second variable is displayed as the column variable. To add margins to the table, we wrap the `addmargins` function around the `table` function.

```
> addmargins(table(Slasher$Survival.char,
Slasher$Gender.char))

           Female Male Sum
  Died        172  228 400
  Survived     50   35  85
  Sum         222  263 485
```

Out of the 222 female characters in slasher films, 172 died, and out of the 263 male characters in slasher films, 228 died. Calculating the conditional distribution of survival by character gender will give us an even clearer picture of the relationship between the two variables.

Deliverable 5: Modify the contingency table in Deliverable 4 to show the conditional distribution of survival by gender.

In general, when calculating a conditional distribution, it's helpful to assign the "by" variable as the second, column variable in the `table` function. As we will see in a later deliverable, this ordering will create the correct display in a side-by-side bar chart. We are interested in the conditional distribution of character survival *by* gender, so we will place `Survived.char` first and `Gender.char` second in the `table` function.

```
> prop.table(
    table(Slasher$Survival.char,Slasher$Gender.char),
    margin = 2
          )*100
              Female        Male
 Died       77.47748    86.69202
 Survived   22.52252    13.30798
```

R prints the conditional distribution of each variable in a contingency table using the `margin` = option. Specifying `margin = 1` generates the distribution of the column variable conditioned on the value of the row variable. Similarly, `margin = 2` generates the distribution of the row variable conditioned on the value of the column variable. Because we want the distribution of survival conditional on gender, we use `margin = 2`.

Only 13% of male compared to 22.5% of female characters survived to the end of the movie.

Deliverable 6: Create a side-by-side bar chart displaying the conditional distribution of survival by character gender.

Because of the way we set up the `prop.table` function when we calculated the conditional distribution of survival by gender, all we have to do is wrap the `barplot` function around the `table` and `prop.table` functions and create the appropriate labels and legend to generate the plot. The second, column variable in the table

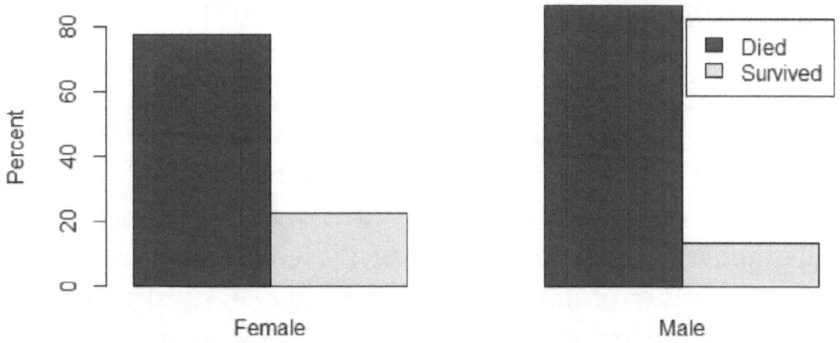

FIGURE 5.1 Character survival in slasher films by gender.

function is used by barplot as the "by" variable on the x-axis and the percent of the first, row variable is graphed on the y-axis.

Type and run the following barplot function in your R Script to create Figure 5.1:

```
barplot(
    prop.table(
         table(Slasher$Survival.char,
               Slasher$Gender.char),
         margin = 2) *100,
     beside = TRUE,
     main = "Character Survival in Slasher Films by
             Gender",
     legend.text = TRUE,
     ylab = "Percent",
         )
```

Deliverable 7: Draw conclusions about the risk of death for male and female characters in slasher films.

Unsurprisingly, the mortality rate for characters in slasher horror films is very high. Male characters were more likely to die than female characters.

STUDENT PROJECT: DETERMINE THE IMPACT PASSENGER AGE AND SEX HAD ON THE LIKELIHOOD A PASSENGER SURVIVED THE *TITANIC* SINKING

In the early hours of April 15, 1912, the unsinkable ship *RMS Titanic* sank when it struck an iceberg, killing more than half of the passengers and crew aboard. The `Titanic.xlsx` dataset contains demographic information for 889 of those passengers as well as a record of whether or not that each passenger survived[2]. Our goal is to determine the impact passenger age and sex had on the likelihood of surviving the *Titanic* sinking.

Project deliverables:

1. Import the `Titanic.xlsx` dataset into RStudio. Use the data dictionary to identify each variable in the dataset as categorical or quantitative. If the variable is categorical, further identify it as ordinal, nominal, or an identifier variable. If the variable is quantitative, identify it as discrete or continuous.
2. Using the `ifelse` function, add a new character variable to the *Titanic* dataset that classifies each passenger as being a child (age less than 18 years) or an adult (age greater than or equal to 18 years). Additionally, add a new variable called `Survived.char` that takes on the value `Did Not Survive` when `Survived == 0` and `Survived` when `Survived == 1`.
3. Create a contingency table showing the joint distribution of survival and age (child vs. adult). Add the table margins to show the marginal distribution of each variable.
4. Modify the contingency table to show the conditional distribution of survival by age (child vs. adult).
5. Draw a side-by-side bar chart to compare the conditional distributions of survival by age category.
6. Create a contingency table showing the joint distribution of survival and sex. Add the table margins to show the marginal distribution of each variable.

7. Modify the contingency table to show the conditional distribution of survival by sex.
8. Draw a side-by-side bar chart to compare the conditional distributions of survival by sex.
9. Summarize your findings about how a passenger's age and sex may have impacted his or her chance of surviving the sinking of the *Titanic.*

REFERENCES

1. Welsh, A. (2010). "On the Perils of Living Dangerously in the Slasher Horror Film: Gender Differences in the Association Between Sexual Activity and Survival," *Sex Roles*, Vol. 62, pp. 762–773.
2. Stanford University. (2016). A *Titanic* Probability. *Titanic* Training Dataset. https://web.stanford.edu/class/archive/cs/cs109/cs109.1166/problem12.html

CHAPTER 6

Linear Regression and Correlation

INTRODUCTION

Scatterplots, linear correlation, and regression allow us to determine the straight-line relationship between two quantitative variables. Scatterplots visualize the relationship between the two variables and give us a general sense of how they vary in relation to one another. A regression line is a mathematical model that uses the value of one variable (called the independent or explanatory variable, interchangeably) to predict the value of the other variable (called the dependent or response variable, interchangeably). Correlation and R^2 measure the strength of the linear relationship modeled by the regression line. They allow us to determine if there is a strong, weak, or no association between the independent and dependent variables.

SCATTERPLOTS AND LINEAR REGRESSION

The first step in linear regression is to determine which variable in the dataset is the explanatory variable and which is the response variable. Usually this is done through careful reading of background information or as you develop your research question. In general terms, the explanatory or independent variable is the one you wish to use to predict the response or dependent variable. It is important to be clear on which variable is which because switching them in the regression model will produce totally different (and incorrect!) results.

Once you have identified the explanatory and response variables, you must check the conditions for modeling the data using linear regression. First, the two variables must be quantitative. Linear regression is not the model to use if you want to examine the relationship between one or more categorical variables. Second, the relationship between the two variables should be linear. Third, there should be no outliers in the data.

The most direct way to test the conditions for linear regression is by creating a scatterplot. A scatterplot graphs each ordered pair (independent variable, dependent variable), allowing you to see if the data points fall more or less in a straight line (with either a positive or negative trend), have a curved (nonlinear) relationship, or have no relationship and look like a cloud of points. Violations of the conditions for linear regression are typically obvious from a scatterplot (and a residual plot), and later we'll discuss what to do if this happens.

Once you've determined that it is appropriate to model your data using linear regression, R makes it very easy to calculate the slope and intercept terms and predict the mean response for a particular value of the independent variable.

THE CORRELATION COEFFICIENT (r) AND THE COEFFICIENT OF DETERMINATION (R^2)

The correlation coefficient (also called the correlation or r, interchangeably) and the coefficient of determination (also called R^2) are related mathematically (R^2 is the square of the correlation) and they measure the strength of the linear relationship between two quantitative variables in slightly different ways. Correlation measures the relationship between the two variables on a scale of –1 to 1 with –1 and 1 being the strongest relationships (negative and positive, respectively) and 0 being no relationship. A correlation of exactly –1 or exactly 1 would mean that all the data points fell in a straight line.

R^2 is measured on a scale of 0 to 1 (recall that the square of a negative number is positive) and is usually expressed as a percent from 0% to 100%. R^2 tells us the percent of variation in the response variable that is explained by differences in the explanatory variable. If R^2 is close to 100%, then knowing what the independent variable is tells us almost everything we need to know to estimate the value of the dependent variable. In contrast, if R^2 is small, then there is missing information in our model that would, if present, allow us to do a better job predicting the dependent variable.

GUIDED PROJECT 1: CAN FORENSIC SCIENTISTS DETERMINE HOW LONG A BODY HAS BEEN BURIED BY HOW MUCH NITROGEN REMAINS IN THE LONG BONES OF THE SKELETON?

Anyone who is a fan of detective TV shows has watched a scene where human remains are discovered and some sort of expert is called in to determine when the person died. But is this science fiction or science fact? Is it possible to use evidence from skeletal remains to determine how long a body has been buried (a decent approximation of how long the person has been dead)?

Researchers sampled long bone material from bodies exhumed from coffin burials in two cemeteries in England. In each case, date of death and burial (and therefore interment time) was known[1]. This data is given in the `Longbones.xlsx` dataset. We wish to determine if there is a relationship between the nitrogen composition of the long bones of the skeleton and the length of time the body was interred.

Project deliverables:

1. Import the `Longbones.xlsx` dataset into RStudio. Use the data dictionary to identify each variable in the dataset as categorical or quantitative. If the variable is categorical, further identify it as ordinal, nominal, or an identifier variable. If the variable is quantitative, identify it as discrete or continuous.

2. Create a scatterplot to display the relationship between long bone nitrogen composition and length of interment. Describe any patterns apparent from the scatterplot.

3. Study authors were concerned that contamination with fuel oil might change the relationship between long bone nitrogen composition and length of interment. On the scatterplot, identify the three burial sites that were contaminated with fuel oil using a different color dot. The study authors decided to remove these observations from analysis. Do you agree?

4. Create a new dataset using only records of remains that were not contaminated by fuel oil. Use this dataset to complete Deliverables 5–10. Are the conditions met to model the relationship using correlation and linear regression?

5. Calculate the correlation between long bone nitrogen composition and length of interment.

6. Determine the regression equation that relates long bone nitrogen composition and length of interment. Interpret the slope and intercept of the line.

7. Create a scatterplot to show the relationship between long bone nitrogen composition and interment time, and annotate it with the regression line and correlation.

8. Create a residual plot and look for potential violations of the conditions for linear regression.

9. Use the regression line to determine the expected mean interment time for a body discovered with a long bone nitrogen composition of 3.71 g/100 g of bone.

10. Calculate R^2 for this relationship. Does long bone nitrogen composition seem to be a good predictor of length of interment?

Deliverable 1: Import the `Longbones.xlsx` dataset into RStudio. Use the data dictionary to identify each variable in the dataset as categorical or quantitative. If the variable is categorical,

further identify it as ordinal, nominal, or an identifier variable. If the variable is quantitative, identify it as discrete or continuous.
The first variable is `Site`, which is a nominal categorical variable identifying from which of two sites the remains were exhumed. The second variable is `Time`, which is a quantitative, continuous variable measuring the time the body was interred in years. The third variable is `Depth`, which is a quantitative, continuous variable measuring the depth at which the body was buried. The fourth variable is `Lime`, which is a nominal, categorical variable indicating if the body was buried with Quicklime. The fifth variable is `Age`, which is a quantitative, continuous variable recording the age at the time of death (though this was unknown in some cases). The sixth variable is `Nitro`, which is a quantitative, continuous variable measuring the nitrogen composition in the femur of the skeleton in g/100 g of bone. The seventh variable is `Oil`, which is a nominal, categorical variable that indicates if the burial site was contaminated with fuel oil.

Deliverable 2: Create a scatterplot to display the relationship between long bone nitrogen composition and length of interment. Describe any patterns apparent from the scatterplot.
We use the `plot` function to create a scatterplot for long bone nitrogen composition and length of interment. The first variable entered in the `plot` function is the independent variable, which will appear on the x-axis, and the second variable is the dependent variable, which will appear on the y-axis. Because we are trying to predict length of interment from long bone nitrogen composition, nitrogen composition is treated as the independent variable and length of interment as the dependent variable. The other plotting statements should be familiar by now. Type and run the following statement to produce Figure 6.1:

```
plot(
      x = Longbones$Nitro,
      y = Longbones$Time,
      main = "Long Bone Nitrogen Composition and
             Length of Interment",
```

```
    Ylim = c(0,100),
    Xlim = c(3,4.5),
    Xlab = "Nitrogen Composition in g per 100 g of
             Bone",
    Ylab = "Length of Interment in Years"
      )
```

The plotted points exhibit a linear trend suggesting a negative linear relationship between long bone nitrogen composition and length of interment. In bodies with longer interment time, there is either greater variability in the nitrogen composition of the long bones or possibly a few outliers.

Deliverable 3: Study authors were concerned that contamination with fuel oil might change the relationship between long bone nitrogen composition and length of interment. On the scatterplot, identify the three burial sites that were contaminated with fuel oil using a different color dot. The study authors decided to remove these observations from analysis. Do you agree?

To color the dots of the burial sites contaminated with fuel oil, we need to create a new variable that assigns the burial sites that were not contaminated to have `Color = 1` and the sites that were

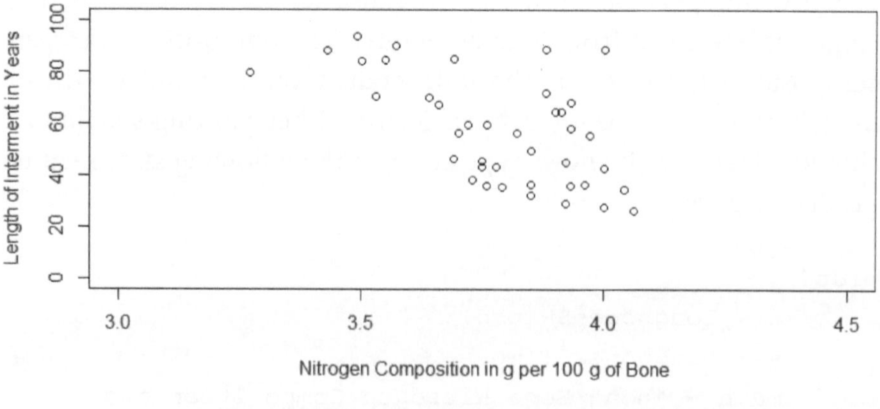

FIGURE 6.1 Long bone nitrogen composition and length of interment.

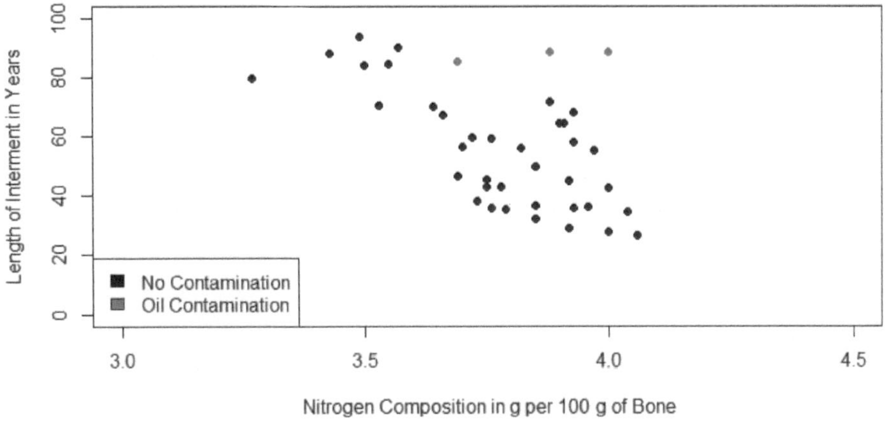

FIGURE 6.2 Long bone nitrogen composition and length of interment.

contaminated to have `Color = 2`. Type and run the following statement in your R Script:

```
Longbones$Color<-Longbones$Oil+1
```

In the plot statement, we determine which colors we wish to use for the contaminated and uncontaminated burial sites (in this case we'll use gray and black, respectively) and in brackets type `[Longbones$Color]` to indicate that is the variable we wish to use to identify which points should be colored differently. Points where `Color = 1` will be printed in black and points where `Color = 2` will be printed in gray.

We can also add a legend to indicate what each color means. For a more thorough discussion of the `legend()` function, see Chapter 2 Guided Project Deliverable 6.

The full statement to type and run in your R Script to produce Figure 6.2 is:

```
plot(
      x = Longbones$Nitro,
      y = Longbones$Time,
      main = "Long Bone Nitrogen Composition and
              Length of Interment",
```

```
        ylim = c(0,100),
        xlim = c(3,4.5),
        xlab = "Nitrogen Composition in g per 100 g of
                Bone",
        ylab = "Length of Interment in Years",
         col = c("black","gray")[Longbones$Color]
         )

legend(
        x = "bottomleft",
        legend = c("No Contamination",
                   "Oil Contamination"),
        fill = c("black","gray")
        )
```

All three oil-contaminated burial sites (appearing in the figure as gray dots) were interred for about 80 years and had between 3.75 and 4.0 g of nitrogen per 100 g of bone remaining in their long bones. Because there are chemical and biological reasons that the oil contamination could alter the natural relationship between long bone nitrogen composition and length of interment, it makes sense to remove those points from the analysis.

Deliverable 4: Create a new dataset using only records of remains that were not contaminated by fuel oil. Use this dataset to complete Deliverables 5–10. Are the conditions met to model the relationship using correlation and linear regression?

We'll create our new dataset using the `subset` function where `Oil == 0`.

```
Longbones.subset <-
     subset(
          x = Longbones,
          subset = (Longbones$Oil == 0)
            )
```

Both variables are quantitative, the scatterplot doesn't show any strong nonlinear shape and we have removed three outliers from the data. All of the conditions for modeling the relationship between long bone nitrogen composition and interment time using linear regression are met.

Deliverable 5: Calculate the correlation between long bone nitrogen composition and length of interment.

The `cor` function is used to calculate the correlation between two variables. It doesn't matter what order they are entered in the function.

```
> cor(Longbones.subset$Nitro,Longbones.subset$Time)
[1] -0.6961346
```

The correlation between long bone nitrogen composition and length of interment is about −0.70, which indicates a moderately strong negative linear relationship.

Deliverable 6: Determine the regression equation that relates long bone nitrogen composition and length of interment. Interpret the slope and intercept of the line.

We use the `lm` function to fit a regression line to the data. The variables (names only, the dataset is specified later in the `lm` function) are entered in the order `y~x`, and, unlike correlation, it is very important that they are entered in that order. Again, because we are trying to predict length of interment from long bone nitrogen composition, length of interment is treated as the y variable and entered first and nitrogen composition is treated as the x variable and entered second. The dataset is named in the `dataset=` entry.

At this point, we'll go ahead and name the regression model because we will need to call it later for other deliverables. To see the results, just type the name of the model.

```
> Longbones.reg<-lm(Time ~ Nitro, data = Longbones.
  subset)
> Longbones.reg
```

```
Call:
lm(formula = Time ~ Nitro, data = Longbones.subset)
Coefficients:
(Intercept)        Nitro
     335.47       -74.26
```

The results return the coefficients of the intercept and the slope. Rewriting them as a model, we have

$$\hat{y} = 335.47 - 74.26\, x$$

The slope term tells us that, for every additional gram of nitrogen per 100 g of bone found in the long bones of a skeleton, we would expect the body to have been interred for about 74.26 fewer years. The intercept tells us what the interment time would be for a skeleton with a long bone nitrogen composition of 0 g per 100 g of bone. Because a nitrogen composition of 0 g per 100 g of bone is outside the range of the data collected, it doesn't make sense to interpret the *y* intercept.

Deliverable 7: Create a scatterplot to show the relationship between long bone nitrogen composition and interment time, and annotate it with the regression line and correlation.

Here is the R code to create the scatterplot using the subset of records that were not contaminated with fuel oil.

```
plot(
     x = Longbones.subset$Nitro,
     y = Longbones.subset$Time,
     main = "Long Bone Nitrogen Composition and
             Length of Interment",
     ylim = c(0,100),
     xlim = c(3,4.5),
     xlab = "Nitrogen Composition in g per 100 g of
             Bone",
     ylab = "Length of Interment in Years"
     )
```

To amend the scatterplot so that it includes both the regression line and correlation, we need to use two additional functions. The first is the `abline` function which will draw the regression line on top of the scatterplot. The `abline` function will print any line on a graph, but in this case we specifically want to print the results of the regression model which we named `Longbones_reg`.

The second function we need is `text`, which will print the formula for the regression equation as well as the correlation. The first two entries in each `text` function are the x and y positions for the center of the textbox on the graph. We will center the top text entry at the (x,y) coordinate (3.3 g/100 g of bone, 20 years) and the second text entry at the (x,y) coordinate (3.3 g/100 g of bone, 10 years).

The `expression` function in each of the text entries allow us to include mathematical symbols in the text added to the graph. For example, `expression` will cause `hat(y)` to be printed on the graph as $\hat{y}$. The syntax to generate other special characters can be found at https://stat.ethz.ch/R-manual/R-patched/library/grDevices/html/plotmath.html

Type the following statements in your R Script and run them after the previous `plot` function to overlay the regression line and text boxes on the scatterplot to produce Figure 6.3:

```
abline(Longbones.reg)
text(
      x = 3.3,
      y = 20,
      expression(hat(y) == 335.47-74.26*x)
      )

text(
      x = 3.3,
      y = 10,
      expression(r == 0.70)
      )
```

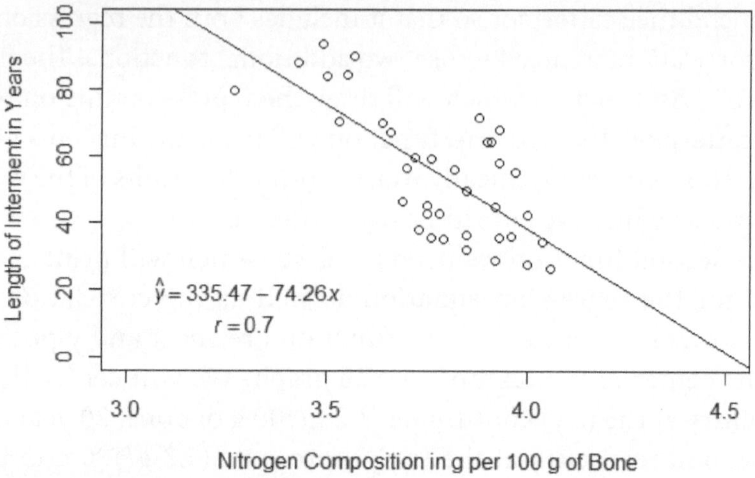

FIGURE 6.3 Long bone nitrogen composition and length of interment.

Deliverable 8: Create a residual plot and look for potential violations of the conditions for linear regression.

When R models a regression line, it computes lots of ancillary data that can be used for other analyses. For example, `Longbones.reg$residuals` is a vector of all the residuals for the `Longbones.reg` regression model.

We can then plot nitrogen composition (the x-variable) on the x-axis against the model residuals by entering `Longbones.reg$residuals` as the y variable.

The `abline` function is also useful here too to add a horizontal line at y = 0 (specified as `h = 0`) on the residual plot. Type and run the following in your R Script to produce Figure 6.4:

```
plot(
      x = Longbones.subset$Nitro,
      y = Longbones.reg$residuals,
      main = "Residual Plot for the Relationship
              Between Long Bone Nitrogen Composition
              and Interment Time",
      xlab = "Nitrogen Composition in g per 100 g of
              Bone",
```

```
    ylab = "Regression Model Residuals"
    )

abline(h=0)
```

There do not appear to be any outliers in the data or any curved pattern that would indicate a nonlinear relationship between long bone nitrogen composition and interment time.

Deliverable 9: Use the regression line to determine the expected mean interment time for a body discovered with a long bone nitrogen composition of 3.71 g/100 g of bone.

R also has a function called `predict` that will calculate $\hat{y}$ for a specified value of x. The first entry in `predict` is the regression model we developed earlier.

The second entry specifies the value of x we wish to model, and it has to be written in a specific way that is a bit different than other functions we've used. The format is `newdata = data.frame(variable name = value)`. This can be a bit confusing because `newdata` should be typed just was written, but

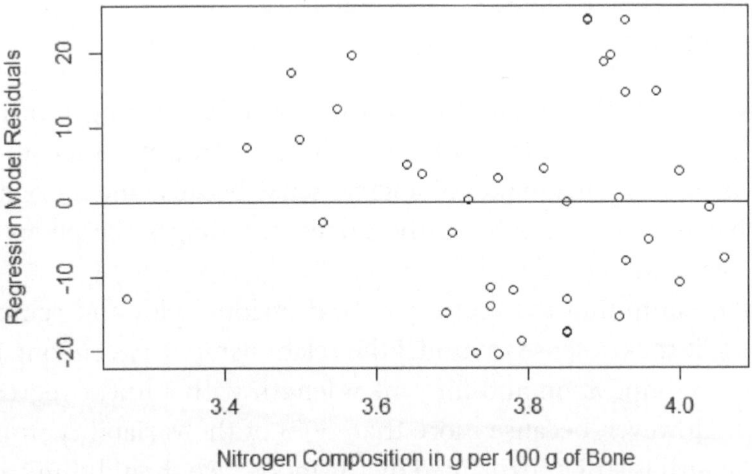

FIGURE 6.4 Residual plot for the relationship between long bone nitrogen composition and interment time.

`variable name` should be replaced with your actual x-variable name and `value` should be replaced with your actual x-variable value.

```
> predict(
      Longbones.reg,
      newdata = data.frame(Nitro = 3.71)
         )
       1
59.98825
```

Based on our regression model, the average interment time for a body discovered with a long bone nitrogen composition of 3.71 g/100 g of bone is about 60 years.

Deliverable 10: Calculate R^2 for this relationship. Does long bone nitrogen composition seem to be a good predictor of length of interment?

The `summary` function provides a lot of information about a regression model and can be used to determine the value of R^2, the coefficient of determination.

```
> summary(Longbones.reg)$r.squared
[1] 0.4846034
```

The R^2 tells us that about 48% of the variability in length of interment can be explained by differences in the nitrogen composition of the long bones. This makes biological sense because the more nitrogen that has decomposed in the bones, the longer the person has been dead and buried.

After examining the scatterplot and residual plot and removing outliers, it makes sense to model the relationship between long bone nitrogen composition and interment length with a linear regression model. However, because more than 50% of the variability in interment length is still unexplained by the model, we should think about expanding our model to include other potential predictors such as age at the time of death (which is, alas, beyond the scope of this book).

GUIDED PROJECT: DETERMINE THE RELATIONSHIP BETWEEN LAND AREA AND THE NUMBER OF PLANT SPECIES LIVING ON THE BRITISH ISLES

How do environmental factors contribute to the diversity of plant life on an island? The `Plants.xlsx` dataset provides information on a number of characteristics of the British Isles (not including Ireland or Britain itself) including latitude, area, and distance from Britain[2,3]. We wish to use linear regression to determine if islands with a larger area are home to a greater number of species of plant life.

Project deliverables:

1. Import the `Plants.xlsx` dataset into RStudio. Use the data dictionary to identify each variable in the dataset as categorical or quantitative. If the variable is categorical, further identify it as ordinal, nominal, or an identifier variable. If the variable is quantitative, identify it as discrete or continuous.
2. Create a scatterplot to display the relationship between island area and number of plant species living in the British Isles.
3. Determine the regression equation that relates island area to the number of plant species.
4. Modify the scatterplot from Deliverable 2 to include the regression line and correlation.
5. Create a residual plot and look for potential violations of the assumptions for linear regression.
6. Does island area seem to be related to the number of plant species living on each of the British Isles? Is linear regression the correct method for modeling this relationship?

Deliverable 1: Import the `Plants.xlsx` dataset into RStudio. Use the data dictionary to identify each variable in the dataset as categorical or quantitative. If the variable is categorical, further

identify it as ordinal, nominal, or an identifier variable. If the variable is quantitative, identify it as discrete or continuous.
The first variable is `Island`, which is a categorical, identifier variable which gives the name of the island. The second variable is `Area`, which is a continuous, quantitative variable measuring the island's area in square kilometers. The third variable is `Elevation`, which is also a continuous, quantitative variable giving the highest elevation on the island in meters. The fourth variable is `Soil`, which is a quantitative, discrete variable counting the number of soil types present on the island. The fifth variable is `Latitude`, which is a quantitative, continuous variable measuring the location of the island in degrees north. The sixth variable is `Distance`, which is a quantitative, continuous variable measuring the distance of the island from Britain in kilometers. The seventh variable is `Species`, which is a quantitative, discrete variable counting the number of plant species living on the island.

Deliverable 2: Create a scatterplot to display the relationship between island area and the number of plant species living in the British Isles.
We plot the scatterplot using the `plot` function. Type and run the following statement in your R Script to generate Figure 6.5:

```
plot(
      x = Plants$Area,
      y = Plants$Species,
      main = "Relationship Between Island Area
              and Number of Plant Species",
      ylab = "Number of Species",
      xlab = "Area (Square Kilometers)"
    )
```

Here we see a different picture from the last example relating long bone nitrogen composition and interment time. In this scatterplot, there seems to be a curved relationship between island area and

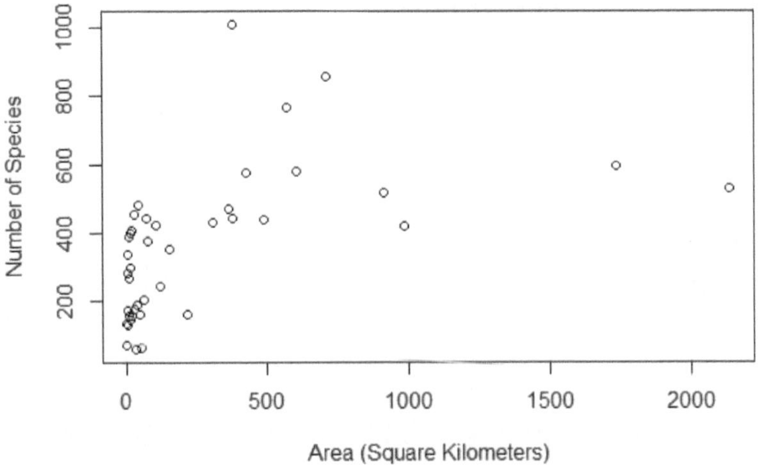

FIGURE 6.5 Relationship between island area and number of plant species.

number of species. There may also be some possible outliers. While area and number of plant species may be related to each other, this scatterplot should make us think that linear regression may not be the right way to model the relationship.

Deliverable 3: Determine the regression equation that relates island area to the number of plant species.

Running the relevant code produces the following output:

```
> reg.plants<-lm(Species ~ Area, data = Plants)
> reg.plants

Call:
lm(formula = Species ~ Area)
Coefficients:
(Intercept)          Area
    295.965         0.238
```

The regression equation for the relationship is $\hat{y} = 295.965 + 0.238x$.

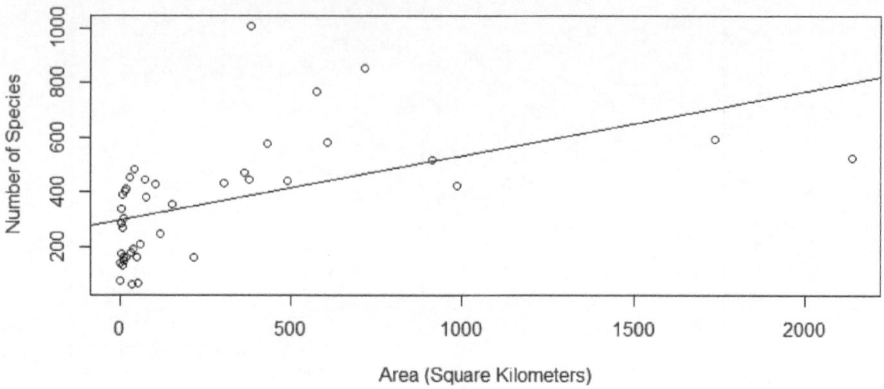

FIGURE 6.6 Relationship between island area and number of plant species.

Deliverable 4: Modify the scatterplot from Deliverable 2 to include the regression line.

Type and run the following statements in your R Script to generate Figure 6.6:

```
plot(
      x = Plants$Area,
      y = Plants$Species,
      main = "Relationship Between Island Area
              and Number of Plant Species",
      ylab = "Number of Species",
      xlab = "Area (Square Kilometers)"
    )

abline(reg.plants)
```

When we add the regression line to the scatterplot we more clearly see the violation of the regression assumptions. There appears to be a curved relationship in the data, and the regression line doesn't model it well.

Deliverable 5: Create a residual plot and look for potential violations of the assumptions for linear regression.

Type and run the following statements in your R Script to generate Figure 6.7:

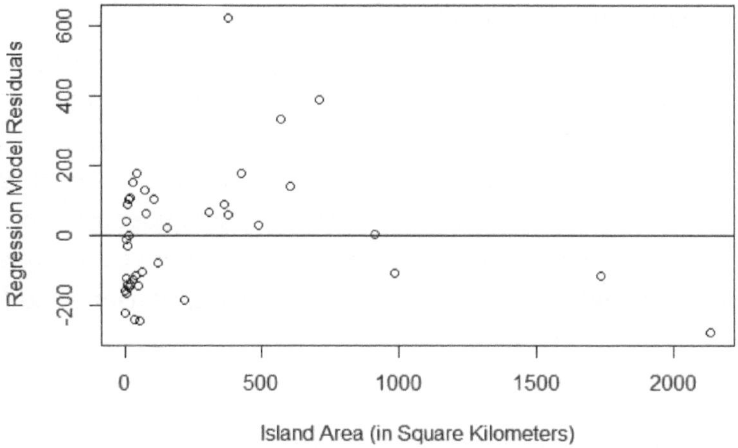

FIGURE 6.7 Residual plot for the relationship between island area and number of plant species.

```
plot(
      x = Plants$Area,
      y = reg.plants$residuals,
      main = "Residual Plot for the Relationship
              Between Island Area and Number of Plant
              Species",
      xlab = "Island Area (in Square Kilometers)",
      ylab = "Regression Model Residuals"
    )

abline(h=0)
```

The residual plot shows the nonconstant variance even more clearly. This is definitely a violation of the assumptions for linear regression.

Deliverable 6: Does island area seem to be related to the number of plant species living on each of the British Isles? Is linear regression the correct method for modeling this relationship?

Yes, island area does seem to be related to the number of plant species living on each island. Islands with larger areas have a larger and more variable number of species living on them. However; linear regression is not the right tool for modeling this relationship.

Hitting a roadblock like this is not unusual in data analysis, and it isn't an indication that you can't find an answer to your research question. While they are beyond the scope of this book, there are transformations (like squaring or taking the square root or logarithm of the dependent variable) that can bring a curved relationship into a straight line or improve nonconstant variance so linear regression can still be used.

STUDENT PROJECT: IS HEAD SIZE RELATED TO BRAIN WEIGHT IN HEALTHY ADULT HUMANS?

The `Brainhead.xlsx` dataset provides information on 237 individuals who were subject to postmortem examination at the Middlesex Hospital in London around the turn of the 20th century[4]. Study authors used cadavers to see if a relationship between brain weight and other more easily measured physiological characterizes such as age, sex, and head size could be determined. The end goal was to develop a way to estimate a person's brain size while they were still alive (as the living aren't keen on having their brains taken out and weighed). We wish to determine the relationship between head size and brain weight in healthy human adults.

Project deliverables:

1. Import the `Brainhead.xlsx` dataset into RStudio. Review the data dictionary to identify each variable in the dataset as categorical or quantitative. If the variable is categorical, further identify it as ordinal, nominal, or an identifier variable. If the variable is quantitative, identify it as discrete or continuous.
2. Create a scatterplot to display the relationship between head size and brain weight. Describe any patters apparent from the scatterplot.
3. Calculate the correlation between head size and brain weight.

4. Determine the regression equation that relates head size and brain weight. Interpret the slope and intercept of the line.
5. Modify the scatterplot from Deliverable 2 to include the regression line and correlation.
6. Create a residual plot and look for potential violations of the assumptions for linear regression.
7. Use the regression line to determine the expected brain weight for a person with a head size of 4000 cubic cm.
8. Calculate R^2 for this relationship. Does head size seem to be a good predictor of brain weight?

REFERENCES

1. Jarvis, D. R. (1997). "Nitrogen Levels in Long Bones from Coffin Burials Interred for Periods of 26–90 Years," *Forensic Science International*, Vol. 85, pp. 199–208.
2. Johnson, M. P. and Simberloff, D. S. (1974). "Environmental Determinants of Island Species Numbers in the British Isles," *Journal of Biogeography*, Vol. 1, pp. 149–154.
3. McCoy, E. D. and Connor, E. F. (1976). "Environmental Determinants of Island Species Number in the British Isles: A Reconsideration," *Journal of Biogeography*, Vol. 3, pp. 381–382.
4. Gladstone, R. J. (1905). "A Study of the Relations of the Brain to the Size of the Head," *Biometrika*, Vol. 4, pp. 105–123.

CHAPTER 7

Random Sampling

INTRODUCTION

Over the course of the next several chapters, we will learn strategies for making statistical inference – in other words, for drawing conclusions about a population from a sample. The first assumption when performing statistical inference is that your sample(s) are representative of the population(s) of interest. However, selecting truly representative samples is possibly the most difficult (and important!) factor in a statistical analysis. If a statistic calculated from a sample is not an accurate estimate of the population parameter, the study is of little (or no!) value.

When working with humans in particular, there are many practical considerations involved in identifying a population of interest and ensuring that your selected study participants actually participate. While these are very important – and you can read more about them in your main statistics textbook – this chapter will focus on the computational aspects of sample selection.

Using R or other software (or a good, old-fashioned random number table) is essential when doing sample selection because the one thing that unites all good samples is that they are selected using randomization and not human choice. We will learn to implement three sampling methods: simple random sampling, stratified sampling, and cluster sampling, and discuss the benefits of each. Large surveys like the ones that occur before national elections often use very complex sampling schemes (multilayer combinations of cluster and stratified sampling) to try to ensure that the samples they select are unbiased.

SAMPLING METHODS

The simplest sampling method is the simple random sample (SRS). In an SRS, each individual in the population has an equal chance of being selected and each sample of the same size has an equal chance of being selected. The SRS is considered the "gold standard" of sampling methods and works very well for lab rats and machine parts but can have some drawbacks when applied to more complex populations.

For example, simple random sampling can – by chance – fail to select any observations from small but important subgroups in the population. Let's say you believe that older age (90+ years) is an important feature of your population. An SRS – again, just by chance – may not randomly select anyone older than age 90 because that age group makes up a small fraction of the overall population. Stratified sampling is an alternative sampling method that can address this potential problem. In stratified sampling, important population strata (such as age group) are defined, and an SRS is selected from each stratum, thus ensuring that members of each strata are represented in the final sample.

Another potential problem with the SRS is that a population can be clustered in a way that makes it impractical to randomly sample members of each cluster. For example, if you wanted to compare two methods for teaching reading to first graders, it wouldn't be practical to have a first-grade class where a couple students were learning to read using one method, a few were learning to read using the other method and the rest were not part of the study at all. Instead, it would make more sense to randomize entire first-grade classes to either of the two methods or leave them out of the study completely. In cluster sampling, important population clusters are identified and entire clusters are randomly selected to be part of the sample.

GUIDED PROJECT 1: COMPARE DIFFERENT SAMPLING METHODS FOR ESTIMATING THE MEAN FINISHING TIME AT THE DISNEY MARATHON IN 2020

More than 14,000 people finished the 2020 Disney Marathon held on January 12. The results by age and gender group are included in the `Disney.xlsx` dataset[1]. We wish to select a simple random sample and a stratified sample of race participants and compare the mean finishing time in those samples to the mean finishing time for all race participants.

Project deliverables:

1. Import the `Disney.xlsx` dataset into RStudio. Use the data dictionary to identify each variable in the dataset as categorical or quantitative. If the variable is categorical, further identify it as ordinal, nominal, or an identifier variable. If the variable is quantitative, identify it as discrete or continuous.
2. Calculate the mean finishing time for all 2020 Disney Marathon race participants.
3. Determine the relative frequency of marathon finishers in each age-gender group.
4. Create a side-by-side box plot of finishing time by age-gender group.
5. Select a simple random sample of 75 marathon participants and calculate the mean finishing time.
6. Determine the relative frequency of marathon finishers in each age-gender group in the sample. Compare the age-gender distribution in the sample to the age-gender distribution in the entire population of finishers.
7. Select a stratified random sample that includes 0.5% of each age-sex group, and calculate the mean finishing time in this sample.
8. Compare the mean finishing time for the two samples to the mean finishing time for the entire population of finishers.

Deliverable 1: Import the `Disney.xlsx` dataset into RStudio. Use the data dictionary to identify each variable in the dataset as categorical or quantitative. If the variable is categorical, further identify it as ordinal, nominal, or an identifier variable. If the variable is quantitative, identify it as discrete or continuous.

The first variable is `ID`, which is a categorical identifier variable used to identify each runner in the 2020 Disney Marathon. The second variable is `gender`, which is a nominal, categorical variable that identifies the participant's gender. The third variable is `age`, which is a continuous, quantitative variable that identifies the participant's age in years. The fourth variable is `group`, which is an ordinal, categorical variable that places each individual in an age-gender group. The fifth variable is `time`, which is a quantitative, continuous variable that lists the participant's finishing time in hours.

Deliverable 2: Calculate the mean finishing time for all 2020 Disney Marathon race participants.

Run the following `mean` function to calculate the mean finishing time:

```
> mean(Disney$time)
[1] 6.062126
```

The mean finishing time for all participants was 6.06 hours.

Deliverable 3: Determine the relative frequency of marathon finishers in each age-gender group.

Run the following `prop.table` function to calculate the percent of individuals in each age-gender group:

```
> prop.table(table(Disney$group)) * 100

F18-24          F25-29          F30-34          F35-39
3.7998015       8.2588969       8.6558911       9.7405359
F40-44          F45-49          F50-54          F55-59
8.0249539       6.5433149       4.0904580       2.5946406
```

```
F60-64          F65+            M18-24          M25-29
1.1626258       0.5033319       2.1834680       4.4165603
M30-34          M35-39          M40-44          M45-49
6.8977740       7.8406352       7.8689919       6.0612505
M50-54          M55-59          M60-64          M65+
5.1183893       3.0058131       2.0487736       1.1838934
```

Men and women between the ages of 25 and 45 made up the largest proportion of participants. There were many fewer participants who were younger than 25 or older than 60.

Deliverable 4: Create a side-by-side box plot of finishing time by age-gender group.

In previous analyses, we used box plots to display and describe the distribution of a single continuous variable. However, it can also be useful to print multiple box plots side-by-side so you can compare their distributions.

Type and run the following `boxplot` function in an R Script to generate Figure 7.1. The argument `Disney$time ~ Disney$group` instructs R to create a box plot of finishing time for every age-gender group. The `las = 2` statement causes the box plot labels to be rotated 90 degrees, and `xlab = ' '` will cause the x axis to remain blank (printed x-axis text gets in the way of the rotated box plot labels in this figure).

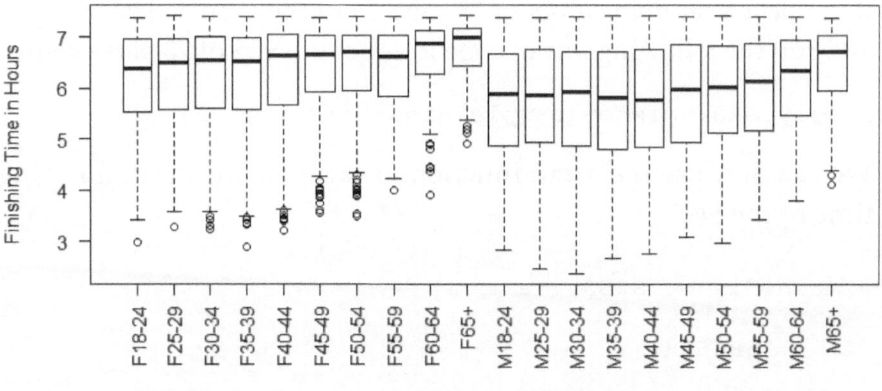

FIGURE 7.1 Disney Marathon 2020 finishing times by age and gender group.

```
boxplot(
    Disney$time ~ Disney$group,
    main = 'Disney Marathon 2020 Finishing Times
            by Age and Gender Group',
    ylab = 'Finishing Time in Hours',
    xlab = '  ',
    las = 2
    )
```

The median finishing time is longer and there are fewer fast outliers in the older age-gender groups.

Deliverable 5: Select a simple random sample of 75 marathon participants and calculate the mean finishing time.

When working with functions that make random selections from a population, it can be helpful to use the `set.seed` function. The `set.seed` function sets the seed of R's random number generator so the results of the random selection will be the same if it is performed multiple times or by multiple individuals. It doesn't matter what number you set the seed to, as long as the seed numbers match, you will get the same results from the random process.

```
set.seed(1234)
```

The following `sample` function (below inside the brackets) instructs R to select a random subset of 75 rows from the Disney dataset. Type and run the following line of code to select the SRS of 75 participants:

```
Disney.SRS<-Disney[sample(nrow(Disney), 75), ]
```

We can now use the mean function to calculate the mean finishing time in the SRS.

```
> mean(Disney.SRS$time)

[1] 5.668652
```

The mean finishing time for the SRS of 75 participants was 5.67 hours.

Deliverable 6: Determine the relative frequency of marathon finishers in each age-gender group in the sample. Compare the age-gender distribution in the sample to the age-gender distribution in the entire population of finishers.

Running the following `prop.table` function will calculate the percent of individuals in the sample in each age-gender group:

```
> prop.table(table(Disney.SRS$group))*100

F18-24      F25-29      F30-34      F35-39
9.333333    6.666667    9.333333    6.666667
F40-44      F45-49      F50-54
5.333333    6.666667    6.666667
F55-59      F60-64      F65+        M18-24
2.666667    1.333333    1.333333    4.000000
M25-29      M30-34      M35-39
2.666667    8.000000    8.000000
M40-44      M45-49      M50-54      M55-59
5.333333    8.000000    4.000000    4.000000
```

We see that the SRS – just by chance – did not select any male participants who were 60 years or older.

Deliverable 7: Select a stratified random sample that includes 0.5% of each age-sex group, and calculate the mean finishing time in this sample.

The `stratified` function is not included automatically when you download R and RStudio, so you will have to install and load it using the `install.packages` and `library` functions below. Type and run the following two statements in your R Script:

```
install.packages("splitstackshape")
library("splitstackshape")
```

We also need to set the seed again so our results are reproducible.

```
set.seed(1234)
```

Type and run the following statements in your R Script to select the stratified random sample:

```
Disney.stratified<-
      stratified(
                 Disney,
                 group = "group",
                 size = .005
                )
```

The first entry in the stratified function is the dataset that we want to sample from. The `group =` argument indicates the name of the variable to stratify on. The `size =` argument is the size of each sample to be selected. If `size =` a number between 0 and 1, a random sample of that proportion will be selected from each stratum. If `size =` a positive integer, R will select a random sample of that size from each stratum. Because `size = 0.005`, the function will randomly select 0.5% of each stratum to be included in the sample. We chose this proportion so the stratified sample and the SRS would be similar in size.

Calculate the mean finishing time for the stratified sample.

```
> mean(Disney.stratified$time)
[1] 6.09234
```

The mean finishing time for individuals in the stratified random sample was 6.09 hours.

Deliverable 8: Compare the mean finishing time for the two samples to the mean finishing time for the entire population of finishers. To review, the mean finishing time in the population and in the two samples was:

```
> mean(Disney$time)
[1] 6.062126

> mean(Disney.SRS$time)
[1] 5.668652
```

```
> mean(Disney.stratified$time)
[1] 6.09234
```

Both samples give an estimate of the mean finishing time that are similar to the population mean. The mean finishing time estimated using the sample random sample is about 6.5% below the population mean and the mean time estimated using the stratified random sample is about 0.5% above the population mean. In this example, the estimate using the stratified sample is closer to the true population mean finishing time than the estimate using the SRS.

Keep in mind, though, that these are the results of a single set of random samples. If we generated two new random samples starting with a different seed, it is possible that the SRS – by chance – would be a better estimate of the population mean than the stratified sample.

GUIDED PROJECT 2: COMPARE DIFFERENT SAMPLING METHODS FOR ESTIMATING THE MEAN HEIGHT OF NATIONAL HOCKEY LEAGUE (NHL) PLAYERS IN 2013–2014

Sports fans often love to know all about their favorite players including their heights, weights, and ages, and professional leagues frequently collect and publish this information. A roster of players from the 2013–2014 NHL season is included in the `NHL.xlsx` dataset[2]. Because players are clustered in different locations by the team they play for, we wish to determine if a cluster sample (using team as the clustering variable) produces a good estimate of the mean height of all NHL players for the 2013–2014 season.

Project deliverables:

1. Import the `NHL.xlsx` dataset into RStudio. Use the data dictionary to identify each variable in the dataset as categorical or quantitative. If the variable is categorical, further identify it as ordinal, nominal, or an identifier variable. If the variable is quantitative, identify it as discrete or continuous.
2. Calculate the mean height of all NHL players in 2013–2014.

3. Determine the mean player height for each team.
4. Select a simple random sample of 75 NHL players, and calculate the mean height.
5. Select a cluster sample with three randomly selected NHL teams. Calculate the mean height of the players in the cluster sample.
6. Compare the mean player height in the two samples to the mean height of the entire population of NHL players during the 2013–2014 season.

Deliverable 1: Import the `NHL.xlsx` dataset into RStudio. Use the data dictionary to identify each variable in the dataset as categorical or quantitative. If the variable is categorical, further identify it as ordinal, nominal, or an identifier variable. If the variable is quantitative, identify it as discrete or continuous.

The first variable is `Team`, which is a nominal, categorical variable. The second variable is `Player`, which is a categorical, identifier variable identifying each unique player. The third variable is `Position`, which is a nominal, categorical variable indicating each player's position. The fourth, fifth, and sixth variables are `Age`, `Weight`, and `Height`, which are all quantitative, continuous variables.

Deliverable 2: Calculate the mean height of all NHL players in 2013–2014.

Run the following `mean` function to calculate the mean height of all NHL players during the 2013–2014 season:

```
> mean(NHL$Height)
[1] 73.26081
```

The mean height of all NHL players in the 2013–2014 season was 73.26 inches.

Deliverable 3: Determine the mean playerheight for each team.

The `aggregate` function splits data into subsets (typically based on the values of a categorical variable), computes summary statistics for each subset, and returns the results in a table. In this case, we wish to split the data into teams (our subsets), calculate the mean player height for each team, and return the results.

Type and run the `aggregate` function in your R Script to produce the following results:

```
> aggregate(
     NHL$Height ~ NHL$Team,
     FUN = mean
           )
                  NHL$Team NHL$Height
1            Anaheim Ducks   73.79167
2          Arizona Coyotes   73.45833
3            Boston Bruins   73.43478
4           Buffalo Sabres   72.72000
5           Calgary Flames   73.37037
6      Carolina Hurricanes   73.78261
7       Chicago Blackhawks   73.87500
8       Colorado Avalanche   72.96154
9    Columbus Blue Jackets   72.75000
10            Dallas Stars   73.53846
11       Detroit Red Wings   73.18519
12         Edmonton Oilers   73.26087
13        Florida Panthers   72.69565
14       Los Angeles Kings   73.70833
15          Minnesota Wild   73.81481
16      Montreal Canadiens   73.17391
17     Nashville Predators   73.04167
18       New Jersey Devils   72.96429
19        New York Rangers   73.84000
20         Ottawa Senators   73.24000
21     Philadelphia Flyers   72.44000
22     Pittsburgh Penguins   73.03846
23         San Jose Sharks   73.40000
24         St. Louis Blues   73.08333
25     Tampa Bay Lightning   73.08000
```

```
26    Toronto Maple Leafs   74.26087
27      Vancouver Canucks   73.22727
28    Washington Capitals   72.47826
29          Winnipeg Jets   73.08000
```

The mean player height for each team is quite similar – right around 73 inches.

Deliverable 4: Select a simple random sample of 75 NHL players and calculate the mean height.

Using the example in Deliverable 5 in Guided Project 1, we can select a random sample of 75 players from the NHL dataset using the following commands:

```
set.seed(1234)
NHL.SRS<-NHL[sample(nrow(NHL), 75), ]
```

The resulting mean height is

```
> mean(NHL.SRS$Height)
[1] 73.16
```

The mean height for the SRS of 75 NHL players was 73.16 inches.

Deliverable 5: Select a cluster sample from the NHL dataset that includes three randomly selected teams. Calculate the mean height of the players in the cluster sample.

The `cluster` function is also not included automatically when you download R and RStudio, so you will have to install and load it using the `install.packages` and `library` functions below.

```
install.packages("sampling")
library("sampling")
```

Set the seed for the random number generator to ensure your results can be reproduced.

```
set.seed(1234)
```

Type and run the following code in your R Script to select the teams that will be included in the cluster sample. Note that it will take one more step to actually assemble the cluster sample dataset.

```
cl<-cluster(
         NHL,
         clustername = c("Team"),
         size = 3,
         method = c("srswor")
            )
```

The first entry in the `cluster` function is the dataset we want to sample from. The `clustername = c("Team"),` argument indicates that we want to use `Team` as the clustering variable. The `size = 3` argument specifies that we want to select three teams. Finally, the `method =` argument indicates the method used to select the sample of clusters. We wish to use the `"srswor"` method, which means simple random sample without replacement.

Unlike the `stratified` function, the `cluster` function does not return all the variable values along with the rows of data selected for the sample. We have to take one more step and use the `getdata` function, which will return all of the rows of `NHL` that appear in `cl` (the players on the three teams selected by the `cluster` function).

```
NHL.cluster<-getdata(NHL,cl)
```

Finally, we can calculate the mean height of the players in the cluster sample.

```
> mean(NHL.cluster$Height)
[1] 73.30435
```

The mean height of NHL players in the cluster sample is 73.3 inches.

Deliverable 6: Compare the mean player height in the two samples to the mean height in the entire population of NHL players during the 2013–2014 season.
To review, the mean height in the population and in the two samples was:

```
> mean(NHL$Height)
[1] 73.26081

> mean(NHL.SRS$Height)
[1] 73.16

> mean(NHL.cluster$Height)
[1] 73.30435
```

Both samples give an estimate of the mean player height that are very similar to the population mean. The mean height estimated using the SRS is about 0.1% below the population mean and the mean height estimated using the cluster sample is about 0.06% above the population mean. Because players are naturally clustered in locations by teams, using a cluster sample to estimate mean player height might be preferable – from an efficiency standpoint – to using an SRS.

STUDENT PROJECT: COMPARE DIFFERENT SAMPLING METHODS FOR ESTIMATING THE MEAN HEIGHT OF A PREMIER LEAGUE PLAYER IN 2014–2015

Yes, more about professional athletes. A roster of players from the 2014–2015 Premier league season is included in the `EPL.xlsx` dataset[3]. We wish to estimate the mean height of all players using an SRS, a sample stratified by position played, and a cluster sample using team as the clustering variable and compare these estimates to the population mean.

Project deliverables:

1. Import the `EPL.xlsx` dataset into RStudio. Use the data dictionary to identify each variable in the dataset as categorical or quantitative. If the variable is categorical, further identify it

as ordinal, nominal, or an identifier variable. If the variable is quantitative, identify it as discrete or continuous.

2. Calculate the mean height of all Premier League players in 2014–2015.
3. Select a simple random sample of 50 Premier League players and calculate the mean height.
4. Draw a side-by-side box plot displaying player height by position played.
5. Create a relative frequency table showing the percent of Premier League players at each position.
6. Select a stratified sample that includes 10% of the players at each position, and calculate the mean player height for the sample.
7. Calculate the mean player height for each Premier League team.
8. Select a cluster sample with three randomly selected Premier League teams, and calculate the mean player height.
9. Compare the mean player height for the three samples to the mean height for the entire population of Premier League players during the 2014–2015 season.

REFERENCES

1. Track Shack. (2020). 2020 Disney Marathon Race Results. https://www.trackshackresults.com/disneysports/results/wdw/wdw20/mar_results.php
2. National Hockey League. (2013). National Hockey League Player Statistics. www.nhl.com
3. Premier League Football. (2014). Premier League Football Player Statistics. www.premierleague.com

CHAPTER 8

Inference About a Population Mean

INTRODUCTION

Up to this point, we've focused our analyses on describing the data we have on hand. We determined the casualty rate for 45 companies from Ohio during the US Civil War, modeled the relationship between the amount of nitrogen remaining in a skeleton buried in an English cemetery to how long it had been interred, and calculated the survival rate for adults and children on the *Titanic*.

However, we are now ready to make the leap from describing data for a small set of individuals to doing statistical inference. We do this in two primary ways: by conducting hypothesis tests and creating confidence intervals.

HYPOTHESIS TESTS

A statistical hypothesis test allows us to weigh evidence for or against an existing claim. When we do statistical inference about a mean, our null hypothesis is that the population mean is equal to a particular value. This value may have been determined from another study or may just be conventional wisdom or other educated guess. We then pose an alternative hypothesis – that the mean is greater than, less than, or simply not equal to the value proposed in the null hypothesis.

Many students find setting up statistical hypotheses and drawing conclusions about those hypotheses based on a *p*-value to be counterintuitive at first. This would be a good point to stop and review the chapter(s) in your main statistics book if you still have lingering questions about the mechanics of hypothesis testing.

CONFIDENCE INTERVALS

A confidence interval allows us to make an educated guess with a predetermined level of confidence (often 95%) about the true value of the population mean. The confidence interval formula uses the sample mean, sample standard deviation, and sample size – along with a parameter that allows us to control how confident we want to be – to propose a range of possible values for the population mean. We believe with a certain level of confidence that the population mean is somewhere in that range.

There is a certain amount of controversy in the statistical world over whether it is "better" to make statistical decisions based on the results of hypothesis testing or confidence intervals. A common criticism of hypothesis testing is that produces a "thumbs up" or "thumbs down" decision based on an arbitrary cut point (the significance level). Is there a practical difference between a *p*-value of 0.051 and a *p*-value of 0.049? Maybe not, and yet they would produce different conclusions at the 0.05 significance level. Confidence intervals, on the other hand, produce an estimate of the population mean and leave it to the reader to determine if there is "enough" evidence to reject the null hypothesis. In the author's opinion, there is value (both pedagogical and from a statistical decision-making standpoint) in both types of statistical inference; therefore, both are illustrated in the following projects.

ASSUMPTIONS FOR INFERENCE ABOUT A POPULATION MEAN

Hypothesis tests for a population mean are conducted using a one-sample t-test, and confidence intervals for a population mean make use of the t-distribution. The assumptions are the same for both.

First, the data should be a simple random sample from a larger population. This tends to be a very hard assumption to meet (see Chapter 7 for more discussion about representative sampling), so in practice, we wish to check that the observations are independent of each other (i.e. there are not multiple observations from the same individual), that they are randomly selected, and that they are as representative of the population as possible.

The second assumption is that the observations come from a population that has a Normal distribution or that the sample size is large enough (greater than about 40). Thanks to the Central Limit Theorem, t-procedures still work very well for large non-Normal samples; however, the smaller the sample size, the more important it is that the data be roughly Normally distributed.

GUIDED PROJECT: IS 98.6°F (37.0°C) ACTUALLY THE NORMAL HUMAN BODY TEMPERATURE?

Everyone knows that 98.6°F (37.0°C) is the normal human body temperature. But is that actually correct, and – come to think of it – how does *everyone* know that in the first place?

A German physician named Carl Reinhold August Wunderlich is generally credited with originating this idea, which was based on – reportedly – more than one million axillary temperature readings taken from 25,000 subjects and was published in his 1868 book *Das Verhalten der Eigenwärme in Krankheiten* (which translates to *The Behavior of the Self-Warmth in Diseases*). But was he correct? History tells that his thermometer was a foot long and took 20 minutes to determine a subject's temperature. For a measure that is used so often to determine general health, it would be a good idea to use modern instruments to confirm or refute his results.

In 1992, three physicians from the University of Maryland School of Medicine set out to do just that, measuring body temperatures for 223 healthy men and women aged 18–40 one to four times a day for three consecutive days using an electronic digital thermometer. The mean body temperature was computed for each individual, and this summary measure is recorded in the `Bodytemp.xlsx` dataset[1]. We wish to test the null hypothesis that the true mean healthy human body temperature is 98.6°F against the alternative hypothesis that the true mean healthy human body temperature is not 98.6°F.

Project deliverables:

1. Import the `Bodytemp.xlsx` dataset into RStudio. Use the data dictionary to identify each variable in the dataset as categorical or quantitative. If the variable is categorical, further identify it as ordinal, nominal, or an identifier variable. If the variable is quantitative, identify it as discrete or continuous.
2. Identify the null and alternative hypotheses we wish to test.
3. Identify the appropriate statistical test for these hypotheses. Verify that the assumptions for using that test are met.
4. Determine the distribution of the sample mean under the null hypothesis.
5. Conduct the hypothesis test and report your conclusion at the alpha = 0.05 significance level.
6. Create and interpret a 95% confidence interval for the true mean healthy human body temperature.

Deliverable 1: Import the `Bodytemp.xlsx` dataset into RStudio. Review the data dictionary to identify each variable in the dataset as categorical or quantitative. If the variable is categorical, further identify it as ordinal, nominal, or an identifier variable. If the variable is quantitative, identify it as discrete or continuous.

There are only two variables in the dataset. The first is `ID`, which is a categorical identifier variable used to identify each participant in the study. The second is `Body_temp`, which is a continuous, quantitative variable.

Deliverable 2: Identify the null and alternative hypotheses we wish to test.

The null hypothesis always a statement about the status quo or commonly accepted value of the mean. The alternative hypothesis is a statement of what we are trying to prove. Therefore:

H_0: $\mu = 98.6°F$
H_a: $\mu \neq 98.6°F$

Deliverable 3: Identify the appropriate statistical test for these hypotheses. Verify that the assumptions for using that test are met.

Because we are making inference about a single population mean, as long as the assumptions are met, the appropriate statistical test is a t-test.

The first assumption is that the data is a simple random sample from the larger population. It is stated in the dataset description that the 223 adults in the sample were selected randomly; therefore, this condition is met.

The second assumption is that observations come from a population that has a Normal distribution or that the sample size is large enough (greater than about 40). The sample size is definitely large enough to meet this assumption, and we can also verify that the data appears to come from a population with a Normal distribution by making a histogram and/or a Normal QQ plot.

The qqnorm function will automatically generate a Normal QQ plot. If the points on the Normal QQ plot create a straight diagonal line from the lower left corner to the upper right corner of the plot, it indicates the Normal model is a good fit for the data (see your course textbook for more details on the mechanics of the Normal QQ plot). Type and run the following statement to produce Figure 8.1:

```
qqnorm(Bodytemp$Body_temp)
```

The commands for creating the histogram shown in Figure 8.2 should be familiar at this point.

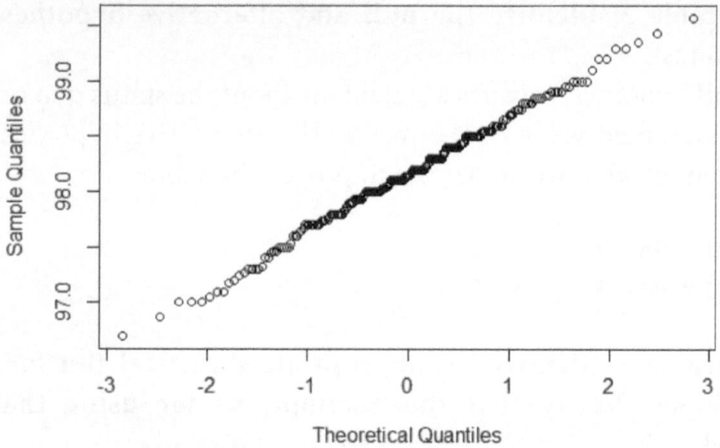

FIGURE 8.1 Normal Q-Q plot.

```
hist(
    Bodytemp$Body_temp,
    main="Body Temperature",
    xlab="Degrees F"
    )
```

Both the Normal QQ plot and the histogram show strong evidence that healthy human body temperature can be described using a

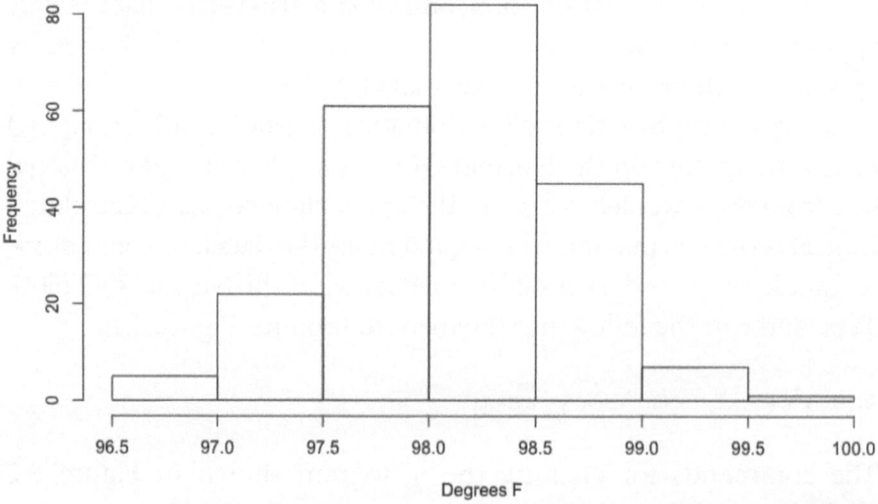

FIGURE 8.2 Body temperature.

Normal distribution. We can proceed feeling confident that a t-test is the correct procedure for testing our hypotheses.

Deliverable 4: Determine the distribution of the sample mean under the null hypothesis.

Under the null hypothesis, the sample mean has a t-distribution with n–1 degrees of freedom. The mean is equal to the mean under the null hypothesis. We'll assign this as `m0`.

```
m0 <- 98.6
```

The standard error is equal to $\frac{s}{\sqrt{n}}$, which we will need to calculate from the dataset. One quick way to find n, the sample size, is by recognizing that it is equal to the number of rows in the dataset, which we can find using the `nrow` function.

```
n <- nrow(Bodytemp)
> n
[1] 223
```

Next, we need to calculate s, the standard deviation of `Body_temp` using the `sd` function.

```
stddev <- sd(Bodytemp$Body_temp)
> stddev
[1] 0.5237612
```

Finally, we calculate the standard error by dividing the standard deviation by the square root of the sample size.

```
se <- stddev/sqrt(n)
> se
[1] 0.03507364
```

Under the null hypothesis and subject to the assumptions we checked earlier, the sample mean has a t-distribution with 222 degrees of freedom. The mean is 98.6°F and the standard error is 0.04°F.

Deliverable 5: Conduct the hypothesis test and report your conclusion at the alpha = 0.05 significance level.

We can find the *p*-value for this hypothesis test in two ways. First, we can do it the "old-fashioned way" by calculating a t-statistic based on our sample and using the `pt` function to determine the *p*-value. The formula to calculate the t-statistic is

$$t = \frac{\bar{x} - \mu_0}{s / \sqrt{n}}$$

We determined the values of μ_0 (the mean under the null hypothesis), s, and $\sqrt{n}$ in Deliverable 4, so all that is left is to find the sample mean, $\bar{x}$, and compute the t statistic.

```
xbar <- mean(Bodytemp$Body_temp)
```

```
> xbar
[1] 98.15919
```

```
t.stat<-(xbar-m0)/se
```

```
> t.stat
[1] -12.56805
```

Now that we have the t-statistic (one, we should note, that is very far from 0), we can calculate the *p*-value using the `pt` function. The `pt` function takes two inputs, the t-statistic and the degrees of freedom and returns the probability of seeing a value equal to or smaller than the t-statistic for a t-distribution with the same degrees of freedom. By default, `pt` returns the left tail or "less than" probability (R calls this the lower tail).

Because we have a two-sided (not equal to) alternative hypothesis, we need to calculate the probability of a t-statistic equal to or smaller than −12.57 and equal to or greater than 12.57. Fortunately, because of the symmetry of the t-distribution, this is as simple as multiplying the results of the `pt` function by 2.

```
> pt(
      t.stat,
      df = 222
      ) * 2
[1] 1.025648e-27
```

The resulting *p*-value is a very small number expressed in scientific notation. The e–27 is an abbreviation for "$\times 10^{-27}$". Therefore, our conclusion is to reject the null hypothesis at the 0.05 significance level and conclude that the mean healthy human body temperature is not 98.6°F.

We can also obtain the same results using R's built-in `t.test` function. The `t.test` function takes the variable of interest as the first entry and the value of the mean under the null hypothesis as the second entry. The default alternative hypothesis is two-sized or "not equal to". Note, however, that you can add a third entry to change the alternative to "greater than" or "less than" using `alternative = c("greater than")` or `alternative = c("less than")`, respectively. In your R Script, type and run the `t.test` function to produce the following results:

```
> t.test(
      Bodytemp$Body_temp,
      mu = 98.6
         )
      One Sample t-test

data:  Bodytemp$Body_temp
t = -12.568, df = 222, p-value < 2.2e-16
alternative hypothesis: true mean is not equal to 98.6
95 percent confidence interval:
 98.09007 98.22831
sample estimates:
mean of x
 98.15919
```

We see that we get the same (or very similar) values for the t-statistic, degrees of freedom, and *p*-value as we did using the `pt` function.

This confirms our decision to reject the null hypothesis at the alpha = 0.05 level and conclude that the mean healthy human body temperature is not 98.6°F.

Deliverable 6: Create and interpret a 95% confidence interval for the true mean healthy human body temperature.
Similarly, there are two ways to calculate a 95% confidence interval for the true mean healthy human body temperature. First, we can do it "the old-fashioned way" by filling quantities into the confidence interval formula $\bar{x} \pm t^* s/\sqrt{n}$, and second, we can use the results of the `t.test` function we used earlier.

The only value we still need to calculate the 95% confidence interval is the critical value, t^*, which we can find using the `qt` function. The `qt` function returns the t value (which we are calling t^*) that corresponds to a specified area below that t value. For a 95% confidence interval, we want the value of t^* such that 95% (0.95 when expressed as a decimal) of the data in our sampling distribution is between $\pm t^*$. That means, because of the symmetry of the t-distribution, 2.5% (or 0.025 expressed as a decimal) of the data in the sampling distribution is below $-t^*$ and 2.5% is above t^*. The only other value we need in the `qt` function is the number of degrees of freedom, which we have already determined to be 222.

```
t.star <- qt(0.025,222)
```

```
> t.star
[1] -1.970707
```

Now we can plug all the quantities into the confidence interval formula to find the estimate of the true mean healthy human body temperature. In your R Script, type and run the following commands to calculate `lower.cl` and `upper.cl`:

```
lower.cl <- xbar + t.star * se
```

```
> lower.cl
[1] 98.09007
```

```
upper.cl <- xbar - t.star * se
```

```
> upper.cl
[1] 98.22831
```

Our calculations are confirmed by the results of the `t.test` function from the previous deliverable (copied in part).

```
95 percent confidence interval:
98.09007 98.22831
```

We are 95% confident that the true mean healthy human body temperature is between 98.09°F and 98.23°F.

STUDENT PROJECT: HOW STANDARD IS THE CAFFEINE CONTENT IN FOUNTAIN SOFT DRINKS SERVED IN CHAIN RESTAURANTS?

Soft drinks like Coke and Pepsi are manufactured to have a standard caffeine content. For example, a 12-oz serving of Coke has 34 mg of caffeine, and a 12-oz serving of Pepsi has 37.6 mg of caffeine. However, fountain soft drinks are typically mixed in individual restaurant dispensers, so it is more difficult to maintain a standard level of caffeine per serving. In this study, researchers randomly sampled Coke, Diet Coke, Pepsi, and Diet Pepsi at a set of franchise restaurants and measured the caffeine content in 12 oz of each soft drink[2]. The data is found in the `Soda.xlsx` dataset.

Because individuals can be sensitive to caffeine – and because manufacturers are interested in product consistency – we wish to test the null hypothesis that the mean caffeine content in 12 oz of Coke served in franchise restaurants is 34 mg versus the alternative hypothesis that the mean caffeine content in 12 oz of Coke served in franchise restaurants is greater than the 34 mg.

Project deliverables:

1. Import the `Soda.xlsx` dataset into RStudio. Use the data dictionary to identify each variable in the dataset as categorical or quantitative. If the variable is categorical, further identify it as ordinal, nominal, or an identifier variable. If the variable is quantitative, identify it as discrete or continuous. Create a

subset called `Coke` that only includes caffeine measurements from regular (not Diet) Coke.

2. Identify the null and alternative hypotheses we wish to test.
3. Identify the appropriate statistical test for these hypotheses. Verify that the assumptions for using that test are met.
4. Calculate the sampling distribution of the sample mean under the null hypothesis.
5. Conduct the hypothesis test and report your conclusion at the alpha = 0.05 significance level.
6. Create and interpret a 95% confidence interval for the true mean amount of caffeine found in fountain Coke.

REFERENCES

1. Mackowiak, P. A., Wasserman, S. S., and Levine, M. M. (1992). "A Critical Appraisal of 98.6 Degrees F, the Upper Limit of the Normal Body Temperature, and Other Legacies of Carl Reinhold August Wunderlich," *Journal of the American Medical Association*, Vol. 268, pp. 1578–1580.
2. Garand, A. N. and Bell, L. N. (1997). "Caffeine Content of Fountain and Private-Label Store Brand Carbonated Beverages," *Journal of the American Dietetic Association*, Vol. 97, no. 2, pp. 179–182.

CHAPTER 9

Inference About a Population Proportion

INTRODUCTION

Hypothesis tests and confidence intervals for a population proportion are very similar to the analogous statistical procedures for means. In this case, we wish to test if the proportion of individuals in a population is equal to a particular value (the null hypothesis) or if it is greater than, less than, or not equal to that value (the choice of alternative hypotheses). If data from our sample leads us to reject the null hypothesis, we then create a 95% confidence interval to make an educated guess about the true population proportion.

ASSUMPTIONS FOR INFERENCE ABOUT A POPULATION PROPORTION

Hypothesis tests and confidence intervals for a population proportion make use of the z-distribution.

First, the data must meet the independence and randomization assumptions, meaning the data should be randomly selected from and representative of the larger population. As mentioned in the last chapter, in practice, we wish to check that the observations are independent of each other (i.e. there are not multiple observations from the same individual) and that they are randomly selected (as much as reasonably possible).

Second, the sample size should be no larger than 10% of the population size. This may seem contradictory – isn't a larger sample size always better? – however, if the sample is too large, it may inadvertently include observations that are dependent on each other, which would also violate the independence assumption.

Finally, there should be at least 10 successes (defined as 10 occurrences of the event of interest) and 10 failures (defined as 10 times the event of interest did not occur).

GUIDED PROJECT: DETERMINE IF FATAL CAR ACCIDENTS ARE MORE LIKELY TO OCCUR ON FRIDAY THE 13TH THAN OTHER FRIDAYS

The `Friday13.xlsx` dataset contains records of deaths from traffic accidents for Friday the 13th and other Fridays in Finland from 1971 to 1997. Traffic on Friday the 13th made up about 3.12% of traffic on all Fridays in Finland during that time period. If Friday the 13th is like all other Fridays, then we would expect 3.12% of Friday traffic fatalities to occur on Friday the 13th also. On the other hand, if there is some kind of Friday the 13th effect (distraction due to superstitions about the day, for example) then we would expect there to be more fatalities on Friday the 13th than on other Fridays. Good Fridays were excluded from the analysis[1]. We wish to test the null hypothesis that 3.12% of Friday traffic fatalities happened on Friday the 13th versus that alternative hypothesis that more than 3.12% of Friday traffic fatalities happened on Friday the 13th.

Project deliverables:

1. Import the `Friday13.xlsx` dataset into RStudio. Use the data dictionary to identify each variable in the dataset as categorical or quantitative. If the variable is categorical, further identify it as ordinal, nominal, or an identifier variable. If the variable is quantitative, identify it as discrete or continuous.

2. Identify the null and alternative hypotheses we wish to test.

3. Identify the appropriate statistical test for these hypotheses. Verify that the assumptions for using this test are met.
4. Calculate the proportion and percent of car crash fatalities that occurred on a Friday the 13th during this time period in Finland.
5. Determine the distribution of the sample proportion under the null hypothesis.
6. Conduct the hypothesis test and report your conclusion at the alpha = 0.05 significance level.
7. Create a 95% confidence interval for the proportion of car crash fatalities on Friday the 13th.
8. Draw conclusions about whether car crash fatalities are more likely on a Friday the 13th compared to other Fridays.

Deliverable 1: Import the `Friday13.xlsx` dataset into RStudio. Use the data dictionary to identify each variable in the dataset as categorical or quantitative. If the variable is categorical, further identify it as ordinal, nominal, or an identifier variable. If the variable is quantitative, identify it as discrete or continuous.
`Friday13.xlsx` has two variables. The first is `ID`, which is a categorical identifier variable that identifies each unique traffic fatality. The second variable is `Fatality` which is a nominal, categorical variable that is equal to `0` if the fatality occurred on a Friday that was not Friday the 13th and is equal to `1` if the fatality occurred on a Friday the 13th.

Deliverable 2: Identify the null and alternative hypotheses we wish to test.
Based on the dataset description, the null hypothesis is that the proportion of Friday traffic fatalities that occur on Friday the 13th is equal to 3.12% (the same as the proportion of Friday traffic that occurs on the 13th). The alternative hypothesis is that the proportion of Friday traffic fatalities that occur on Friday the 13th is greater

than 3.12% (higher than the proportion of Friday traffic on the 13th). In symbolic notation, this is

$$H_0: p = 0.0312$$
$$H_a: p > 0.0312$$

Deliverable 3: Identify the appropriate statistical test for these hypotheses. Verify that the assumptions for using that test are met. There are four assumptions that must be met to conduct a z-test for one proportion: the independence assumption, the randomization assumption, the 10% assumption, and the success/failure assumption.

First, we test the independence assumption by asking if it reasonable to believe that records of traffic fatalities in the dataset are independent of each other. While there could be some fatalities that were not independent (i.e. two occurred when one vehicle hit another), for the most part, traffic fatalities on Fridays spanning nearly three decades would be expected to be independent of each other.

Second, we need to check the randomization assumption. Were the traffic fatalities in this sample randomly selected from the population of all Friday traffic fatalities? No. The authors have drawn a convenience sample of records from Finland. It is possible that these results would not hold in another country – say a country that had no history of superstition around Friday the 13th. However, we'll proceed with the analysis knowing there are some limits to the conclusions we can draw.

Third, the 10% assumption. The 10% assumption states that the sample size should be no more than 10% of the size of the population. If we consider our population to be all of the passengers in all of the cars operating on Fridays (or at least cars in operation in places with superstitions about Friday the 13th), then our sample is much smaller than 10% of the population.

Finally, the success/failure assumption. As we will see in the next deliverable, there are more than 10 "successes" and 10 "failures" in the dataset.

Other than the random sample condition, which limits our ability to draw broad conclusions, there are no violations of the assumptions needed to conduct a one-proportion z-test. We can proceed with the analysis as described in the deliverables.

Deliverable 4: Calculate the proportion and percent of fatalities that occurred on a Friday the 13th.

In this scenario, a traffic fatality on Friday the 13th is considered a "success" and a traffic fatality that occurred on another Friday is considered a "failure" (highlighting the fact that there is nothing inherently good about "successes" in probability). The sample proportion is calculated by dividing the number of successes by the total number of observations.

We can count the number of successes by using the `sum` function to add up all of the 1s – which indicate Friday the 13th fatalities – in the `Fatality` variable.

```
x <- sum(Friday13$Fatality)
> x
[1] 123
```

Then we can use the `nrow` function to determine the sample size.

```
n <- nrow(Friday13)
> n
[1] 3335
```

We see that the number of successes (123) and the number of failures (3335–123 = 3212) are both greater than 10, meeting the success/failure assumption in the previous deliverable.

The sample proportion is the number of successes divided by the sample size.

```
phat <- x/n
> phat
[1] 0.03688156
```

In our sample, the proportion of Friday traffic fatalities that occurred on Friday the 13th was 0.0369 or 3.69%.

We can also calculate this percent using the `prop.table` and `table` functions.

```
> prop.table(table(Friday13$Fatality)) * 100
        0          1
96.311844  3.688156
```

Of all traffic fatalities that occurred on Fridays, 3.69% occurred on a Friday the 13th.

Deliverable 5: Determine the distribution of the sample proportion under the null hypothesis.

```
p0 <- 0.0312
```

The standard error of the sample proportion is calculated using the formula $\text{SE} = \sqrt{\frac{p_0(1-p_0)}{n}}$

Using RStudio to do the calculations for us, we get

```
se <- sqrt(p0*(1-p0)/n)
> se[1] 0.003009499
```

Under the null hypothesis, subject to the assumptions we checked earlier, the distribution of the sample proportion is Normal with a mean of 0.0312 and a standard error of 0.003.

Deliverable 6: Conduct the hypothesis test and report your conclusion at the alpha = 0.05 significance level.

We can find the p-value for this hypothesis test in several ways. First, we can do it the "old-fashioned way" by calculating a z-statistic based on our sample and using the `pnorm` function to determine the p-value. The formula to calculate the z-statistic is $z = \frac{\hat{p} - p_0}{\sqrt{\frac{p_0(1-p_0)}{n}}}$

Using the quantities we calculated in previous deliverables, we can solve to get the z-statistic.

```
z <-  (phat-p0)/se
> z
[1] 1.887214
```

Now that we have the z-statistic, we can calculate the *p*-value using the `pnorm` function. The `pnorm` function takes the z-statistic as input and returns the probability of seeing a value equal to or smaller than that number. By default, `pnorm` returns the left-tail or "less than" probability (R calls this the lower tail).

Because we have a "greater than" alternative hypothesis, we want to calculate the probability of a z-statistic greater than or equal to 1.88. Specifying the `lower.tail=FALSE` option returns the right-tail or "greater than" probability.

```
> pnorm(
      z,
      lower.tail=FALSE
      )
[1] 0.02956579
```

Another, slightly less old-fashioned way to approach this problem is to use the `pnorm` function specifying the values of p-hat, and the mean and standard error under the null hypothesis as arguments.

```
> pnorm(
      phat,
      p0,
      se,
      lower.tail=FALSE
      )
[1] 0.02956579
```

Another(!) way to find the *p*-value for the hypothesis test is by using the `prop.test` function. `Prop.test` computes a z-test for a single proportion using the number of successes (the first entry), the number of observations (the second entry), and the value of *p* under the null hypothesis (the third entry).

It is also possible to specify the direction of the alternative hypothesis in `prop.test`. The default is "two-sided" (which is the "not equal to" alternative); however, in this case want to use the "greater than" alternative.

```
> prop.test(
      x,
      n,
      p = p0,
      alternative = "greater"
          )

      1-sample proportions test with continuity
          correction

data:  x out of n, null probability p0
X-squared = 3.3761, df = 1, p-value = 0.03307
alternative hypothesis: true p is greater than 0.0312
95 percent confidence interval:
 0.03173855 1.00000000
sample estimates:
         p
0.03688156
```

We see on the second line that the `p-value = 0.03248`, which is very close to the *p*-value we obtained using the `pnorm` function (R uses the "plus 4" method of estimating the sample proportion in the `prop.test` function, which accounts for the slight difference).

All three methods lead to the same conclusion. We reject the null hypothesis at the alpha = 0.05 level and conclude that the rate of traffic fatalities is higher on Friday the 13th than other Fridays.

As a side note, we present multiple methods for approaching the same problem not to confuse or overwhelm students but to emphasize that there are often several ways to get the right answer. We also want to illustrate that the built-in R functions are doing the same calculations students are learning to do by hand and using the z- and t-tables.

Deliverable 8: Create a 95% confidence interval for the proportion of car crash fatalities on Friday the 13th.

Similarly, there are two ways to calculate a 95% confidence interval for the true proportion of Friday the 13th car crashes. First, we can do it "the old-fashioned way" by filling quantities into the confidence interval formula $\hat{p} \pm z^{*}\sqrt{\frac{\hat{p}(1-\hat{p})}{n}}$.

We've already calculated all the values we need to use the confidence interval formula except z^*, which we can find using the qnorm function. The qnorm function returns the z value (which we are calling z^*) that corresponds to a specified area below that z value. For a 95% confidence interval, we want the value of z^* such that 95% (0.95 when expressed as a decimal) of the data in our sampling distribution is between $\pm z^*$. That means, because of the symmetry of the Normal distribution, that 2.5% (or 0.025 expressed as a proportion) of the data in the sampling distribution is below $-z^*$ and 2.5% is above z^*.

```
z.star <- qnorm(0.025)
> z.star
[1] -1.959964

sd.phat <- sqrt(phat*(1-phat)/n)
> sd.phat
[1] 0.003263597
```

Now we can plug in all the quantities into the confidence interval formula to find the estimate of the true proportion of Friday auto fatalities that occur on Friday the 13th. Note that we use proportion instead of p under the null hypothesis to calculate the confidence interval because our estimate of the population proportion isn't tied to any particular hypothesis test.

```
lower.cl <- phat + z.star*sd.phat
> lower.cl
[1] 0.03048503
```

```
upper.cl <- phat - z.star*sd.phat
> upper.cl
[1] 0.04327809
```

Alternately, we can use the `prop.test` function with the default two-sided alternative to get the 95% confidence interval for the true proportion of Friday the 13th fatalities.

```
> prop.test(
      x,
      n,
      p = p0
          )

      1-sample proportions test with continuity
          correction

data:  x out of n, null probability p0
X-squared = 3.3761, df = 1, p-value = 0.06615
alternative hypothesis: true p is not equal to 0.0312
95 percent confidence interval:
 0.03086207 0.04399159
sample estimates:
         p
0.03688156
```

Because the `prop.test` function uses the "plus 4" estimate of proportion, the 95% confidence interval is not exactly the same as the one we calculated by hand. However, they are very similar. We can be 95% confident between 3.1% and 4.4% of Friday traffic fatalities occur on Friday the 13th.

Deliverable 9: Draw conclusions about whether car crash fatalities are more likely to happen on a Friday the 13th compared to other Fridays.

We do conclude that car crash fatalities are more likely to happen on Friday the 13th compared to other Fridays, at least in countries with similar superstitions about Friday the 13th. While driving time on Friday the 13th made up about 3.12% of driving time on all Fridays, about 3.7% of Friday traffic fatalities occurred on Friday the 13th.

STUDENT PROJECT: IS CHOLERA TRANSMITTED BY DRINKING CONTAMINATED WATER?

In the mid-1800s, before microorganisms had been discovered, the prevailing theory was that disease was caused by miasma (bad air). It made a lot of sense. The places where disease was most rampant – the overcrowded homes of destitute city dwellers – smelled absolutely terrible. However, some doctors and scientists were unconvinced. One in particular, a British pioneer of anesthesiology named Dr. John Snow, set out to demonstrate that cholera (a devastating disease in 19th century London where he lived) was transmitted not by the air but by the drinking water.

Snow is better known for his "Ghost Map", which linked an 1854 cholera outbreak in SOHO to a single contaminated drinking pump (and for etherizing Queen Victoria, providing pain relief while she delivered her eighth child). However, this data comes from a study of an 1854 cholera outbreak in South London that was commissioned later by the General Board of Health to confirm Snow's finding that cholera was transmitted by water and not air.

Because of the haphazard development of the sewage system in London in the 1800s, the same general area was often serviced by more than one water company. The South London districts in this study had water supplied by either the Southwark & Vauxhall Company or the Lambeth Company. Although the Lambeth Company drew water from a cleaner portion of the river (the Southwark & Vauxhall Company draw water from a part of the Thames that was heavily polluted with sewage – yuck!), the death rate from cholera during this time for consumers of their water was 0.24%[2]. We wish to test if the death rate from cholera for consumers of Southwark & Vauxhall water was equal to (the null hypothesis) or greater than (the alternative hypothesis) 0.24%.

Project deliverables:

1. Import the `Cholera.xlsx` dataset into RStudio (this is a big dataset and may take a minute). Open the data dictionary to identify each variable in the dataset as categorical or quantitative. If the variable is categorical, further identify it as ordinal, nominal, or an identifier variable. If the variable is quantitative, identify it as discrete or continuous.
2. Create a subset containing only records of individuals who had water provided by the Southwark & Vauxhall Company.
3. Identify the null and alternative hypotheses we wish to test.
4. Identify the appropriate statistical test for these hypotheses. Verify that the assumptions for using that test are met.
5. Calculate the proportion of Southwark & Vauxhall customers who died of cholera.
6. Determine the distribution of the sample proportion under the null hypothesis.
7. Conduct the hypothesis test and report your conclusion at the alpha = 0.05 significance level.
8. Create a 95% confidence interval for the proportion of Southwark & Vauxhall customers who died of cholera.
9. Does the risk of death from cholera seem to depend on water source?

REFERENCES

1. Nayha, S. (2002). "Traffic Deaths and Superstition on Friday the 13th," *American Journal of Psychiatry*, Vol. 159, pp. 2110–2111.
2. Frost, W. H. (1936). Appendix to "*Snow on Cholera*." London: Oxford University Press.

CHAPTER 10

Comparing Two Population Means

INTRODUCTION

Now that we've learned to test hypotheses and calculate confidence intervals for a single population parameter, we wish to extend our inference methods to be able to compare two groups. In this chapter, we will focus on comparing means using two different study designs: two independent samples and matched pairs.

TWO INDEPENDENT SAMPLE DESIGN

In a two independent sample design, random samples are selected from two distinct populations. The mean of each sample is computed, and, using a two-sample t-test, we test if the population means are equal (the null hypothesis) or if one population mean is greater than the other or if they are simply not equal (the possible alternative hypotheses). If the population means are determined not to be equal to each other, we can create a confidence interval to estimate the true difference in the population means.

Assumptions for a two independent sample t-test are similar to those for a one-sample t-test except that they apply to both samples. The first assumption is that each sample is randomly selected from an independent, larger population. The second assumption is that each sample size is large enough (about 40) or that the variable of interest has a Normal distribution in each independent population.

MATCHED PAIRS STUDY DESIGN

In a matched pairs design, the individuals in the two samples are selected to be *dependent* on each other. One classic matched pairs design is a twin study, where one twin experiences one set of experimental conditions and the other twin experiences a different set of experimental conditions. Rather than comparing the overall means for each experimental condition, the interesting result is the difference within each set of twins. The null hypothesis in a matched pairs study is that the *mean difference* is equal to zero and the choice of alternative hypothesis is that the mean difference is greater than, less than, or not equal to zero. Dependent sample studies can also be matched in other ways: for example, a parent and child, two partners in a relationship or the same individual under two sets of experimental conditions.

The assumptions for a matched pairs study design are similar to the assumptions for a t-test for a single population mean but focus on the differences calculated for each study pair. The first assumption is that the data is a simple random sample from the larger population. This can be tricky to think about because, while the two samples are *dependent* on each other, each pair should be chosen to be independent.

The second assumption is that the observations are randomly selected from a population with a Normal distribution or that the sample size is large enough. In a matched pairs design, the sample size is actually the number of pairs of observations. If there are not at least 40 pairs of observations, a histogram of the differences and/or a Normal QQ plot should indicate a roughly Normal distribution.

GUIDED PROJECT 1: CAN FINGERTIP AND UPPER ARM MEASURES OF SYSTOLIC BLOOD PRESSURE BE USED INTERCHANGEABLY?

Accurate measurement of blood pressure is important because elevated blood pressure is associated with serious health outcomes such as heart attack and stroke. While blood pressure is typically measured in the upper arm using a cuff, this is not the only place on the body blood pressure can be taken. As part of a larger study, researchers wished to determine if blood pressure measured in an individual's fingertip was the same as blood pressure measured in the individual's arm. The alternative hypothesis was that one measure was systematically higher than the other. Finger and arm systolic blood pressure measurements for 200 randomly selected individuals are recorded in the `Sysbp.xlsx` dataset[1]. We wish to determine if arm and fingertip measures of systolic blood pressure are equivalent within the same person or if one is systematically higher than the other.

Project deliverables:

1. Import the `Sysbp.xlsx` dataset into RStudio. Use the data dictionary to identify each variable in the dataset as categorical or quantitative. If the variable is categorical, further identify it as ordinal, nominal, or an identifier variable. If the variable is quantitative, identify it as discrete or continuous.

2. Determine the experimental design in this scenario. Is this a matched pairs design or a test for two independent means?

3. Identify the null and alternative hypotheses.

4. Identify the appropriate statistical test for these hypotheses. Verify that the assumptions for using that test are met.

5. Calculate the distribution of the mean difference under the null hypothesis.

6. Conduct your hypothesis test and report your conclusion at the alpha = 0.05 significance level.

7. Create and interpret a 95% confidence interval for the true difference in arm and fingertip systolic blood pressure measurements.
8. Are arm and fingertip measures of systolic blood pressure equivalent within the same person, or is one systematically higher than the other?

Deliverable 1: Import the `Sysbp.xlsx` dataset into RStudio. Use the data dictionary to identify each variable in the dataset as categorical or quantitative. If the variable is categorical, further identify it as ordinal, nominal, or an identifier variable. If the variable is quantitative, identify it as discrete or continuous.
The first variable is `id`, which is a categorical identifier variable that uniquely identifies each study participant. The second variable is `armsys`, which is a continuous, quantitative variable recording the systolic blood pressure measured in the participant's arm. The third variable is `fingsys`, which is a continuous, quantitative variable recording the systolic blood pressure measured in the participant's fingertip.

Deliverable 2: Determine the experimental design in this scenario. Is this a matched pairs design or are we comparing two independent means?
Because we wish to compare the arm and fingertip systolic blood pressure for each individual (the same individual under two experimental conditions), this is a matched pairs design.

Deliverable 3: Identify the null and alternative hypotheses.
In a matched pairs design, the null hypothesis is always that the mean difference for each matched pair of observations is zero (i.e. for each individual, the fingertip and arm systolic blood pressure measures are the same). In our scenario, we don't have any prior conception that systolic blood pressure measured in the arm will be systematically higher than systolic blood pressure measured in the fingertip or vice versa. Therefore, we can write the hypotheses as:

$$H_0 : \mu_d = 0$$
$$H_a : \mu_d \neq 0$$

Deliverable 4: Identify the appropriate statistical test for these hypotheses. Verify that the assumptions for using that test are met.

Because this study has a matched pairs design, the appropriate test for our hypotheses is a matched pairs t-test. Although we have two sets of systolic blood pressure measurements for each individual, we are ultimately doing a one-sample t-test on the difference in arm and fingertip systolic blood pressure for each individual. Therefore, our assumptions are the same as for a one-sample t-test.

The first assumption is that the data is a simple random sample from a larger population. This assumption is satisfied because the dataset description informs us that the 200 participants in the study were randomly selected.

The second assumption is that observations come from a population with a Normal distribution or that the sample size is large enough (about 40). Because there are 200 pairs of systolic blood pressure measurements, this assumption is met.

We can also confirm that the difference between fingertip and arm systolic blood pressure in the population appears to have a Normal distribution by calculating the difference in the two measures for each individual and making a histogram and a Normal QQ plot of the differences. The histogram of the difference in systolic blood pressure measures can be seen in Figure 10.1 and the Normal QQ plot in Figure 10.2.

```
Sysbp$diff <- Sysbp$fingsys - Sysbp$armsys

hist(
      Sysbp$diff,
      main = "Difference in Systolic Blood Pressure
              Measured in the Arm and Fingertip",
```

```
    xlab = "Difference: Fingertip SBP - Arm SBP
            (mmHg)"
    )
> qqnorm(Sysbp$diff)
```

Both the Normal QQ plot and the histogram show strong evidence that the difference in systolic blood pressure measurements can be described using a Normal distribution. We can proceed feeling confident that a matched pairs t-test is the correct procedure for testing our hypotheses.

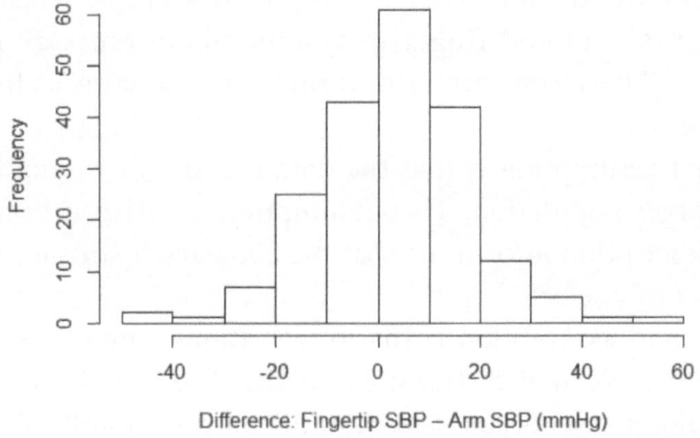

FIGURE 10.1 Difference in systolic blood pressure measured in the arm and fingertip.

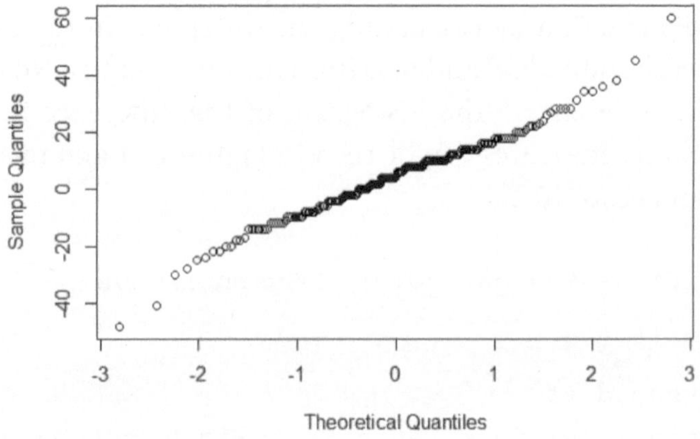

FIGURE 10.2 Normal Q-Q plot.

Deliverable 5: Calculate the distribution of the mean difference under the null hypothesis.

Under the null hypothesis and subject to the assumptions we checked earlier, the mean difference in systolic blood pressure has a t-distribution with n–1 degrees of freedom. The dataset description tells us that there are 200 pairs of systolic blood pressure measurements; therefore, the distribution has 199 degrees of freedom. Under the null hypothesis, the mean difference in systolic blood pressure measured in the fingertip and arm is 0 and the standard error is equal to $\frac{s_d}{\sqrt{n_d}}$, where s_d is the standard deviation of the differences (the `diff` variable we created earlier) and n_d is the number of differences.

```
std.err <- sd(Sysbp$diff)/sqrt(nrow(Sysbp))
> std.err
[1] 1.031411
```

Therefore, the mean difference has a t-distribution with 199 degrees of freedom, a mean of 0 and a standard error of approximately 1.03.

Deliverable 6: Conduct your hypothesis test and report your conclusion at the alpha = 0.05 significance level.

There are three different ways to find the p-value for this hypothesis test. The "old-fashioned way" is to calculate a t-statistic for the mean difference and find the p-value using the `pt` function. Note that, in the t-statistic formula, we subtract the mean difference under the null hypothesis from the mean difference in the sample. Because the mean difference under the null hypothesis is 0, we don't need to include it in our calculations.

```
t <- mean(Sysbp$diff)/std.err
> t
[1] 4.164198
```

Now we use `pt` with a t-statistic of 4.16 and 199 degrees of freedom to obtain the p-value for the hypothesis test. Because the alternative hypothesis is two-sided, we are interested in the probability of seeing a t-statistic as small or smaller than –4.16 or as large or larger than 4.16. The symmetry of the t-distribution means it's easiest to find the *p*-value by calculating the probability of a t-statistic as small or smaller than –4.16 and multiplying the result by 2.

```
> pt(-4.16,199) * 2
[1] 4.731555e-05
```

Another way to find the *p*-value is to use the `t.test` function to test if the mean difference is equal to 0. Note that the default alternative hypothesis is "not equal to" and the default mean under the null hypothesis is 0, so we don't have to specify either in the `t.test` function. Running the following `t.test` in your R Script produces the following output:

```
> t.test(Sysbp$diff)

        One Sample t-test

data:  sysbp$diff
t = 4.1642, df = 199, p-value = 4.652e-05
alternative hypothesis: true mean is not equal to 0
95 percent confidence interval:
 2.261102 6.328898
sample estimates:
mean of x
    4.295
```

Finally, we can use the `t.test` function to compare the difference in blood pressure measurements using the `paired=TRUE` option (without `paired=TRUE` the `t.test` function compares the means of two independent groups). Type and run the following `t.test` function in your R Script:

```
> t.test(
      Sysbp$fingsys,
```

```
        Sysbp$armsys,
        paired=TRUE
            )

Paired t-test

data:  sysbp$fingsys and sysbp$armsys
t = 4.1642, df = 199, p-value = 4.652e-05
alternative hypothesis: true difference in means is
  not equal to 0
95 percent confidence interval:
 2.261102 6.328898
sample estimates:
mean of the differences
                  4.295
```

Regardless of what method we use, our conclusion is that we reject the null hypothesis at the alpha = 0.05 level and conclude that the mean difference in fingertip and arm systolic blood pressure measures is not 0.

Deliverable 7: Create and interpret a 95% confidence interval for the true difference in systolic blood pressure measurements.
The 95% confidence interval for the difference in systolic blood pressure (arm – finger) is shown in the results of the two `t.test` function in Deliverable 6. We are 95% confident that systolic blood pressure measured in an individual's fingertip is between 2.26 and 6.33 mmHg higher than systolic blood pressure measured in the same individual's arm.

Deliverable 8: Are arm and fingertip measures of systolic blood pressure equivalent within the same person, or is one systematically higher than the other?
Based on the results of our hypothesis test and confidence interval, we conclude that arm and fingertip blood pressure measurements are not equivalent within the same person. Fingertip blood pressure seems to be consistently higher than arm blood pressure.

GUIDED PROJECT 2: DETERMINE THE MOST EFFECTIVE MOSQUITO REPELLANT TREATMENT FOR SOLDIERS IN THE INDIAN ARMY

Mosquito nets have traditionally been an important tool to prevent mosquito bites in parts of the world where malaria is endemic. However, it may not be practical for an army that is on the move to set up and carry mosquito nets each night and day. Impregnating soldiers' uniforms with insect repellant solves the mobility problem but also has drawbacks. First, the insect repellant quickly becomes ineffective with repeated washing and ironing and must be frequently reapplied. Second, in hot and humid climates the insect repellant can be absorbed through the skin, and the long-term effects of this exposure are unknown. One compromise is to have soldiers apply patches treated with insect repellant to their clothing. These patches would last longer because they would not be washed or ironed and would not expose the entire body to the insect repellant.

The `Mosquito.xlsx` dataset contains data recorded in an experiment conducted on male soldiers in the Indian Army who were stationed in the Tezpur/Solmara garrison in Northeast India. Thirty soldiers were randomly selected to receive one of five types of mosquito repellant patch. Three of the treatments were a single repellant and two were combinations of two repellants. After giving informed consent, the study participants affixed the patches at predetermined points on their uniforms and research assistants (who were blinded to the type of repellant used) counted the number of times a mosquito landed on each individual in an hour[2]. We wish to determine if there is a difference in the mean number of mosquito landings between soldiers who wore patches with a single repellant and soldiers who wore patches with a combination of two repellants.

Project deliverables:

1. Import the `Mosquito.xlsx` dataset into RStudio. Open the data dictionary and identify each variable in the dataset as categorical or quantitative. If the variable is categorical, further identify it as ordinal, nominal, or an identifier variable. If the variable is quantitative, identify it as discrete or continuous.

2. Determine the experimental design in this scenario. Is this a matched pairs design or a test for two independent means?
3. Identify the null and alternative hypotheses.
4. Identify the appropriate statistical test for these hypotheses. Verify that the assumptions for using that test are met.
5. Conduct your hypothesis test and report your conclusion at the alpha = 0.05 significance level.
6. Create and interpret a 95% confidence interval for the difference in the mean number of mosquito touches between soldiers who wore patches treated with only one kind of mosquito repellant and soldiers who wore patches treated with a combination of two kinds of mosquito repellant.
7. Draw conclusions about the effectiveness of patches with one versus a combination of two types of mosquito repellant.

Deliverable 1: Import the `Mosquito.xlsx` dataset into RStudio. Open the data dictionary and identify each variable in the dataset as categorical or quantitative. If the variable is categorical, further identify it as ordinal, nominal, or an identifier variable. If the variable is quantitative, identify it as discrete or continuous.
The first variable is `ID`, which is a categorical, identifier variable that identifies each unique study participant. The second variable is `Treatment`, which is a nominal, categorical variable indicating which mosquito repellant patch treatment the participant received (one repellant or a combination of two repellants). The third variable is `Mosq_count`, which is a discrete, quantitative variable that gives the number of times the participant was touched by a mosquito in an hour while wearing the repellant patches.

Deliverable 2: Determine the experimental design in this scenario. Is this a matched pairs design or are we comparing two independent means?
Because we wish to compare two independent groups – soldiers wearing patches with one mosquito repellant and soldiers wearing

patches with a combination of two mosquito repellants – we are comparing two independent means. In contrast, if the same soldiers were assigned to wear one-repellant patches and then two-repellant patches, that would be a matched pairs study design.

Deliverable 3: Identify the null and alternative hypotheses.
Let the mean number of mosquito touches on soldiers wearing one-repellant patches be μ_1 and the mean number of mosquito touches on soldiers wearing two-repellant patches be μ_2. In a study design comparing two independent means, the null hypothesis is that the means of the two populations are equal.

$$H_0 : \mu_1 - \mu_2 = 0$$

In this scenario, the alternative hypothesis is that the mean number of mosquito touches will be different depending on the different type of patches being worn.

$$H_a : \mu_1 - \mu_2 \neq 0$$

Note that you also could specify μ_1 to be the mean number of mosquito touches for soldiers wearing two-repellant patches and μ_2 to be the mean number of mosquito touches for soldiers wearing the one-repellant patches.

Deliverable 4: Identify the appropriate statistical test for these hypotheses. Verify that the assumptions for using that test are met.
Because we are comparing the means of two independent groups, we will conduct a two-sample t-test.

The first assumption is that the observations in each group are all independent of each other. In the scenario, we learned that soldiers enrolled in the study were randomly assigned to the two treatments (wearing one-repellant or two-repellant patches). Therefore, this assumption is met.

The second assumption is that the number of mosquito touches in each population has a Normal distribution or that the size of each sample is at least 40. We can check the sample size part of the assumption by making a table of the treatment variable.

```
> table(Mosquito$Treatment)
 0  1
90 60
```

Histograms and QQ plots for the number of mosquito touches in each treatment group show that the distribution of mosquito touches is unimodal but slightly skewed to the right (probably due, in part, to the fact that the number of mosquito touches can't be negative). However, because there are at least 40 observations in each treatment group, the Central Limit Theorem tells us that the distribution of the mean number of mosquito touches in each treatment group will be Normal.

Deliverable 5: Conduct your hypothesis test and report your conclusion at the alpha = 0.05 significance level.

Because calculating the pooled standard error for a two-sample t-test is a bit complicated, it makes more sense to use the `t.test` function rather than trying to find the t-statistic by hand and calculate the *p*-value using the `pt` function.

```
> t.test(
      Mosquito$Mosq_count[Mosquito$Treatment == 0],
      Mosquito$Mosq_count[Mosquito$Treatment == 1]
        )

      Welch Two Sample t-test

data: Mosquito$Mosq_count[Mosquito$Treatment == 0]
  and Mosquito$Mosq_count[Mosquito$Treatment == 1]
t = 3.9539, df = 132.44, p-value = 0.0001246
alternative hypothesis: true difference in means is
  not equal to 0
95 percent confidence interval:
 1.038327 3.117228
```

```
sample estimates:
mean of x mean of y
 8.011111  5.933333
```

Note that the data we need to compare the two means is not stored as two separate variables but is instead identified by `Treatment`. The brackets after `Mosquito$Mosq_count` can be read as "where". So the first entry in the `t.test` function is "the variable `Mosq_count` where `Treatment == 0`" and the second entry is "the variable `Mosq_count` where Treatment `== 1`". In other words, we are comparing the number of mosquito touches for soldiers who were wearing one-repellant patches to the number of mosquito touches for soldiers wearing two-repellant patches. The default difference in the two population means is 0 and the default alternative hypothesis is "`two-sided`".

We can see from the results that the t-statistic is 3.95 and the *p*-value is 0.00012. We reject the null hypothesis at the alpha = 0.05 level and conclude that the mean number of mosquito touches for soldiers wearing one-repellant patches is different than the mean number of mosquito touches for solders wearing two-repellant patches.

Deliverable 6: Create and interpret a 95% confidence interval for the difference in the mean number of mosquito touches between soldiers who wore patches treated with only one kind of mosquito repellant and soldiers who wore patches treated with a combination of two kinds of mosquito repellant.

The 95% confidence interval for the difference in the mean number of mosquito touches is found in the t-test results in the previous deliverable.

```
95 percent confidence interval:
 1.038327 3.117228
```

We are 95% confident that soldiers wearing one-repellant patches will have, on average, 1–3 more mosquito touches than solders wearing two-repellant patches.

Deliverable 7: Draw conclusions about the effectiveness of patches with one versus a combination of two types of mosquito repellant. Results of our hypothesis test and confidence interval indicate that soldiers wearing two-repellant patches have, on average, between one and three fewer mosquito touches in an hour. Because fewer mosquito touches (hopefully!) leads to fewer disease-carrying mosquito bites, the two-repellant patches seem to be more effective.

STUDENT PROJECT 1: DETERMINING LEAD CONCENTRATIONS IN LIP COSMETICS

While lead exposure can cause damage to the kidneys and nervous system of both children and adults, as recently as 2013, a limit for lead in cosmetics had not been defined at either the European or international level. Lead can be absorbed through the skin, so cosmetics are a possible source of daily, population-wide, and often long-term exposure. The `Lead.xlsx` dataset contains records of various lipstick and lip gloss products along with the concentration of lead present in each[3]. We wish to test the hypothesis that lipstick and lip gloss contain different mean amounts of lead.

Project deliverables:

1. Import the `Lead.xlsx` dataset into RStudio. Open the data dictionary and identify each variable in the dataset as categorical or quantitative. If the variable is categorical, further identify it as ordinal, nominal, or an identifier variable. If the variable is quantitative, identify it as discrete or continuous.
2. Determine the experimental design in this scenario. Is this a matched pairs design or are we comparing two independent means?
3. Identify the null and alternative hypotheses.
4. Identify the appropriate statistical test for these hypotheses. Verify that the assumptions for using that test are met.
5. Conduct the hypothesis test and report your conclusion at the alpha = 0.05 significance level.

6. Create and interpret a 95% confidence interval for the difference in the mean lead concentration in lipstick and lip gloss.
7. If you wished to wear lip cosmetics but were concerned about lead exposure, would you choose to wear lipstick or lip gloss?

STUDENT PROJECT 2: ARE THE RESULTS OF A NEW TEST DESIGNED TO ASSESS FAT CONTENT IN HUMAN MILK CONSISTENT WITH THE "GOLD STANDARD" GERBER METHOD

As technology advances, new methods are developed to increase the efficiency and accuracy of laboratory measurements. However, to ensure repeatability and comparability of results, it is important to show that the new measurement yields results that are consistent with the old measurement technique.

The Gerber method of assessing fat content (mainly in dairy products) was developed by a German scientist by that name in the late 19th century. Dr. Gerber's method continues to be used to this day (though it is rivaled in the United States by the similar Babcock method. Stop by Babcock Hall at the University of Wisconsin-Madison for a taste of the Babcock method for determining fat content in ice cream form).

A newer method for determining fat content by enzymic hydrolysis of triglycerides was developed. Both this new "triglyceride" method as well as the Gerber method were used to assess the fat content of 45 samples of human milk. This data is included in the `Milkfat.xlsx` dataset[4]. We wish to determine if there is a systematic difference in the fat content measured by the triglyceride method and the Gerber method.

Project deliverables:

1. Import the `Milkfat.xlsx` dataset into RStudio. Download the data dictionary and identify each variable in the dataset as categorical or quantitative. If the variable is categorical, further identify it as ordinal, nominal, or an identifier variable. If the variable is quantitative, identify it as discrete or continuous.

2. Determine the experimental design in this scenario. Is this a matched pairs design or are we comparing two independent means?
3. Identify the null and alternative hypotheses.
4. Identify the appropriate statistical test for this hypothesis test. Verify that the assumptions for using that test are met.
5. Calculate the distribution of the sample mean under the null hypothesis.
6. Conduct the hypothesis test and report your conclusion at the alpha = 0.05 significance level.
7. Create and interpret a 95% confidence interval for the true mean difference in fat composition measured using the two methods.
8. Draw conclusions about the fat content of human milk as measured by the triglyceride method and the Gerber method.

REFERENCES

1. Bland, J. M. and Altman, D. G. (1995). "Comparing Methods of Measurement: Why Plotting Difference Against Standard Method is Misleading," *Lancet*, Vol. 346, pp. 1085–1087.
2. Bhatnagar, A. and Mehta, V. K. (2007). "Efficacy of Deltamethrin and Cyfluthrin Impregnated Cloth Over Uniform Against Mosquito Bites," *Medical Journal Armed Forces India*, Vol. 63, pp. 120–122.
3. Piccinini, P., Piecha, M., and Torrent, S. F. (2013). "European Survey of the Content in Lead in Lip Products," *Journal of Pharmaceutical and Biomedical Analysis*, Vol. 76, pp. 225–233.
4. Bland, J. M., Altman, D. G. (1999). Measuring Agreement in Method Comparison Studies," *Statistical Methods in Medical Research*, Vol. 8, pp. 135–160.

CHAPTER 11

Comparing Two Population Proportions

INTRODUCTION

In this final chapter, we will extend our inference methods to compare two proportions from independent populations. Similar to a hypothesis test for two independent means, a hypothesis test for two independent proportions compares the proportion of "successes" in two random samples selected from two distinct populations. The null hypothesis is that the population proportions are equal, and the possible alternative hypotheses are that one population proportion is greater than the other or they are simply not equal. If the population proportions are determined not to be equal, we can create a confidence interval to estimate the true difference between them.

Assumptions for doing a z-test for two proportions are similar to those for a z-test for a single proportion except that they apply to both samples. To meet the independence and randomization assumptions, the data in each sample should be randomly selected from each of two distinct larger populations. The second assumption is that each sample size should be no larger than 10% of the corresponding population size. Last, there should be at least 10 successes (defined as 10 occurrences of the event of interest) and 10 failures (defined as 10 times the event of interest did not happen) in each of the samples.

STUDENT PROJECT: ARE FEMALE CHARACTERS IN SLASHER FILMS MORE LIKELY TO BE MURDERED THAN MALE CHARACTERS?

Slasher movies thrill audiences by portraying a lone antagonist (typically male) who attacks innocent victims with extreme violence and without apparent motive. However, this exciting (if gory) subgenre of horror film is criticized by those who view the violence as being used to "punish" female characters who engage in sexual activity during the film. To test this claim, study authors randomly sampled 50 North American slasher films released from 1960 to 2009 and coded the 485 characters appearing in them as being male or female, involved in sexual activity or not, and if they survived the film or not[1]. The data appears in the `Slasher.xlsx` dataset. In this project, we are going to answer a slightly simpler question: Is the proportion of female victims in slasher films the same or different than the proportion of male victims?

Project deliverables:

1. Import the `Slasher.xlsx` dataset into RStudio. Review the data dictionary to identify each variable in the dataset as categorical or quantitative. If the variable is categorical, further identify it as ordinal, nominal, or an identifier variable. If the variable is quantitative, identify it as discrete or continuous.

2. Use the `ifelse` function to create a new variable called `Survival.char` that takes on the value `Survived` when the character survived and `Died` when the character died. Similarly, create another new variable called `Gender.char` that takes on the values `Female` and `Male` for female and male characters, respectively.

3. Calculate the marginal and joint distributions of `Gender` and `Survival`. Identify the number of observations in each independent group (male and female characters) and the number and proportion of individuals in each group who died during the film.

4. What are the null and alternative hypotheses we wish to test? What statistical test should be used for these hypotheses?

5. Verify that the assumptions needed to test the hypotheses in Deliverable 4 are met.

6. Conduct the hypothesis test in Deliverable 5 and report your conclusion at the alpha = 0.05 level.

7. Calculate a 95% confidence interval for the difference in the proportion of male and female slasher movie fatalities.

Deliverable 1: Import the `Slasher.xlsx` dataset into RStudio. Use the data dictionary to identify each variable in the dataset as categorical or quantitative. If the variable is categorical, further identify it as ordinal, nominal, or an identifier variable. If the variable is quantitative, identify it as discrete or continuous. The first variable in the dataset is `ID`, which is a categorical identifier variable that identifies each character uniquely. The second variable is `Gender`, which is a nominal categorical variable identifying each character as being male or female. The third variable is `Activity`, which is a nominal categorical variable indicating if the character was depicted engaging in sexual activity. The fourth variable is `Survival`, which is a nominal categorical variable indicating if the character survived the film or not.

Deliverable 2: Use the `ifelse` function to create a new variable called `Survival.char` that takes on the value `Survived` when the character survived and `Died` when the character died. Similarly, create another new variable called `Gender.char` that takes on the values `female` and `male` for female and male characters, respectively.

Type and run the following statements in your R Script to create `Survival.char` and `Gender.char`:

```
Slasher$Survival.char <-
     ifelse(
          test = (Slasher$Survival == 0 ),
          yes = 'Died',
          no = 'Survived'
            )


Slasher$Gender.char <-
     ifelse(
          test = (Slasher$Gender == 0 ),
          yes = 'Male',
          no ='Female'
            )
```

Deliverable 3: Calculate the marginal and joint distributions of `Gender` and `Survival`. Identify the number of observations in each independent group (male and female characters) and the number and proportion of individuals in each group who died during the film.

We can calculate the joint and marginal distributions of `Survival` and `Gender` using the `table` function with `addmargins`. See Chapter 5 if you need a review of these functions.

```
> addmargins(table(Slasher$Survival.char,
Slasher$Gender.char))

          Female Male Sum
  Died       172  228 400
  Survived    50   35  85
  Sum        222  263 485
```

Out of the sample of 222 female characters, 172 died during the movie, and out of the 263 male characters, 228 died during the movie.

Deliverable 4: What are the null and alternative hypotheses we wish to test? What statistical test should be used for these hypotheses?

We wish to test the null hypothesis that the proportion of female characters who die during slasher films is the same as the proportion of male characters who die during slasher films. The alternative hypothesis is that these proportions are different. In symbols:

$$H_0 : p_{\text{male}} = p_{\text{female}}$$
$$H_a : p_{\text{male}} \neq p_{\text{female}}$$

Deliverable 5: Verify that the assumptions needed to test the hypotheses in Deliverable 4 are met.

The independence and randomization assumptions are met because the 50 slasher films included in the study were randomly selected from all North American slasher films released from 1960 to 2009. We assume that the characters in the randomly selected films are representative of all male and female characters in slasher films. The 10% assumption states that the sample size should be no more than 10% of the size of the population. As long as there were more than 2220 female and 2630 male slasher film characters between 1960 and 2009, this assumption is met. Finally, we see that there are at least 10 victims and 10 survivors among both the male and female characters, so the sample size assumption is met.

Deliverable 6: Conduct the hypothesis test in Deliverable 5 and report your conclusion at the alpha = 0.05 level.

Similar to when we conducted a z-test for one proportion, we can use the `prop.test` function to compute a hypothesis test for two independent proportions. Because we have two sets of "successes" (one for each sample), and two sample sizes, our entries for `x =` and `n =` in the `prop.test` function are vectors. We can see that `x = 228` deaths for the `n = 263` male characters and `x = 172` deaths for the `n = 222` female characters.

```
> prop.test(
     x = c(228, 172),
     n = c(263, 222)
           )

2-sample test for equality of proportions with
  continuity
     correction
data:  c(228, 172) out of c(263, 222)
X-squared = 6.4485, df = 1, p-value = 0.0111
alternative hypothesis: two.sided
95 percent confidence interval:
 0.01940169 0.16488907
sample estimates:
   prop 1    prop 2
0.8669202 0.7747748
```

We reject the null hypothesis at the 0.05 significance level and conclude that the proportion of male and female survivors is not equal.

Deliverable 7: Calculate a 95% confidence interval for the difference in the proportion of male and female slasher movie fatalities. From Deliverable 6:

```
95 percent confidence interval:
 0.01940169 0.16488907
```

For ease of interpretation, we can express the confidence interval above in percents rather than proportions: 1.9%–16.5% (rounded). We are 95% confident that the percent of male characters who die in slasher films is between 1.9 and 16.5 percentage points higher than the percent of female characters who die in slasher films.

STUDENT PROJECT: IS CHOLERA TRANSMITTED BY DRINKING CONTAMINATED WATER?

In the mid-1800s, before microorganisms had been discovered, the prevailing theory was that disease was caused by miasma (bad air). It made a lot of sense. The places where disease was most rampant – the overcrowded homes of destitute city dwellers – smelled absolutely terrible. However, some doctors and scientists were unconvinced. One in particular, a British pioneer of anesthesiology named Dr. John Snow, set out to demonstrate that cholera (a devastating disease in 19th century London where he lived) was transmitted not by the air but by the drinking water.

Snow is better known for his "Ghost Map", which linked an 1854 cholera outbreak in SOHO to a single contaminated drinking pump (and for etherizing Queen Victoria, providing pain relief while she delivered her eighth child). However, this data comes from a study of an 1854 cholera outbreak in South London that was commissioned later by the General Board of Health to confirm Snow's finding that cholera was transmitted by water and not air.

Because of the haphazard development of the sewerage system in London in the 1800s, the same general area was often serviced by more than one water company. The South London districts in this study had water supplied by either the Southwark & Vauxhall Company or the Lambeth Company[2]. In the Chapter 9 Student Project, we slightly simplified this problem and posed it as a hypothesis test for a single proportion. Now we wish to test if the proportion of deaths from cholera for consumers of Southwark & Vauxhall water was the same as or different from the proportion of deaths from cholera for consumers of water from the Lambeth Company.

Project deliverables:

1. Import the `Cholera.xlsx` dataset into RStudio (this is a big dataset and may take a minute). Open the data dictionary to identify each variable in the dataset as categorical or quantitative. If the variable is categorical, further identify it as ordinal, nominal, or an identifier variable. If the variable is quantitative, identify it as discrete or continuous.

2. Use the `ifelse` function to create a new variable called `Survival.char` that takes on the value `Survived` if the individual did not die of cholera and `Died` otherwise. Similarly,

create another new variable called `Company.char` that identifies the drinking water source as `Southwark & Vauxhall` or `Lambeth`.

3. Calculate the marginal and joint distributions of drinking water source and survival. Identify the number of observations in each independent group (Southwark & Vauxhall customers and Lambeth customers) and the number and proportion of individuals in each group who survived.
4. What are the null and alternative hypotheses we wish to test? What statistical test should be used for these hypotheses?
5. Verify that the assumptions needed to test the hypotheses in Deliverable 4 are met.
6. Conduct the hypothesis test in Deliverable 5 and report your conclusion at the alpha = 0.05 level.
7. Calculate a 95% confidence interval for the difference in the proportion of cholera deaths for the two different drinking water sources.

REFERENCES

1. Welsh, A. (2010). "On the Perils of Living Dangerously in the Slasher Horror Film: Gender Differences in the Association Between Sexual Activity and Survival," *Sex Roles*, Vol. 62, pp. 762–773.
2. Frost, W. H. (1936). Appendix to "*Snow on Cholera*". London: Oxford University Press.

Student Project

Does Brain Weight Differ by Age in Healthy Adult Humans?

The `Brainhead.xlsx` dataset provides information on 237 individuals who were subject to postmortem examination at the Middlesex Hospital in London around the turn of the 20th century[1]. Study authors used cadavers to see if a relationship between brain weight and other more easily measured physiological characterizes such as age, sex, and head size could be determined. The end goal was to develop a way to estimate a person's brain size while they were still alive (as the living aren't keen on having their brains taken out and weighed). We wish to determine if there is a relationship between age and brain weight in healthy human adults.

Project deliverables:

1. Import the `Brainhead.xlsx` dataset into RStudio. Review the data dictionary to identify each variable in the dataset as categorical or quantitative. If the variable is categorical, further identify it as ordinal, nominal, or an identifier variable. If the variable is quantitative, identify it as discrete or continuous.
2. Create a histogram of brain weight and calculate the appropriate summary measures to describe the distribution.
3. Display the distribution of age graphically.
4. Describe the distribution of age with a numerical summary.

5. Draw side-by-side box plots illustrating the distribution of brain weight by age.

6. Calculate and compare the mean and standard deviation of brain weight by age.

7. Describe the hypothesis test you would use to test for a statistically significant difference in brain weight by age.

8. Identify the appropriate statistical test for your hypotheses in Deliverable 7, and determine if the assumptions for using this test are met.

9. Test for a statistically significant difference in brain weight by age at the 0.05 level.

10. Calculate a 95% confidence interval for the difference in the mean brain weight for older and younger individuals.

11. Summarize your results about the relationship of age and brain weight in healthy adults.

REFERENCE

1. Gladstone, R. J. (1905). "A Study of the Relations of the Brain to the Size of the Head," *Biometrika*, Vol. 4, pp. 105–123.

Student Project

Preventing Acute Mountain Sickness with Ginkgo Biloba and Acetazolamide

Acute mountain sickness (AMS) is a common concern for mountain climbers who ascend higher than 2000 m. Characterized by headache, lightheadedness, fatigue, nausea, and insomnia, AMS is caused by a failure to adapt to the acute hypobaric hypoxia experienced at high altitudes. The drug acetazolamide has been used effectively to treat AMS; however, it has a variety of unpleasant side effects that can reduce compliance to taking it. Previous studies suggested that the herbal supplement ginkgo biloba might also be used to prevent AMS without side effects. To test this hypothesis, healthy western volunteers who were hiking Mt. Everest were randomized to one of four treatments: placebo, ginkgo biloba only, acetazolamide only or ginkgo biloba and acetazolamide[1]. Treatment group as well as incidence of AMS and incidence of headache for the 487 individuals who completed the experiment are presented in `Treckers.xlsx`. We wish to determine if ginkgo biloba is as effective in preventing AMS as acetazolamide.

Project deliverables

1. Import the `Treckers.xlsx` dataset into RStudio. Open the data dictionary to identify each variable in the dataset as

categorical or quantitative. If the variable is categorical, further identify it as ordinal, nominal, or an identifier variable. If the variable is quantitative, identify it as discrete or continuous.

2. Create a subset that contains records of participants who were randomized to take ginkgo biloba only and acetazolamide only. Hint: use `subset = (Treckers$Trt == 2 | Treckers$Trt ==3)` in your R code. We will use this subset to complete the rest of the deliverables.
3. Create a new variable called `Trt.char` that takes on the value `Ginkgo Biloba` or `Acetazolamide` for individuals who were assigned to those treatments. Create another new variable called `AMS.char` that takes on the value `Yes` for participants who developed AMS and `No` for participants who did not develop AMS.
4. What number and proportion of hikers developed AMS?
5. Calculate the joint and marginal distributions of treatment and AMS.
6. Determine the conditional distribution of the incidence of AMS by treatment.
7. Display the results of Deliverable 6 in a side-by-side bar chart.
8. What is the appropriate test to determine if the proportion of individuals who develop AMS while taking acetazolamide is the same as the proportion who develop AMS while taking ginkgo biloba? Verify that the assumptions for using this test are met.
9. Write the hypotheses for the test you identified in Deliverable 8.
10. Conduct the hypothesis test and report your conclusion at the 0.05 significance level.

11. Create a 95% confidence interval for the difference in the proportion of AMS cases for those taking ginkgo biloba compared to those taking acetazolamide.

12. Summarize your conclusions about the effectiveness of ginkgo biloba and acetazolamide as treatments for AMS.

REFERENCE

1. Gertsch, J. H., Basnyat, B., Johnson, E. W., Onopa, J., and Holck, P. S. (2005). "Randomized, Double-Blind Placebo Controlled Comparison of Ginkgo Biloba and Acetazolamide for Prevention of Acute Mountain Sickness Among Himalayan Trekkers: the Prevention of High Altitude Illness Trial," *British Medical Journal*, Vol. 328, p. 797.

Student Project

What Factors Influence Mammal Sleep Patterns?

All mammals sleep. As any college student who has pulled an all-nighter knows, going without sleep or trying to function on too little sleep has a host of deleterious effects. But for something that is so clearly physiologically important, there is a great variety in sleep needs throughout the animal kingdom from animals that seem never to sleep to those who that seem never to wake (ahem, cats). Researchers recorded data on sleep duration as well as a set of ecological and constitutional variables for a selection of mammal species[1]. This data appears in the `Sleep.xlsx` dataset. We wish to examine the relationship between dreaming and nondreaming sleep time in this set of mammal species.

Project deliverables:

1. Import the `Sleep.xlsx` dataset into RStudio. Open the data dictionary to identify each variable in the dataset as categorical or quantitative. If the variable is categorical, further identify it as ordinal, nominal, or an identifier variable. If the variable is quantitative, identify it as discrete or continuous.
2. Display and describe the distribution of total sleep for the mammal species in the dataset.
3. Plot the relationship between nondreaming and dreaming sleep. Do animals who spend more time in dreaming sleep also

spend more time in nondreaming sleep or does dreaming sleep decrease as nondreaming sleep increases?

4. What is the appropriate method to model the relationship between time spent in nondreaming sleep and time spent in dreaming sleep? Verify that the assumptions for using this method are met.
5. Create a model to predict time spent in dreaming sleep from time spent in nondreaming sleep.
6. Calculate and interpret the correlation and R^2 describing the relationship between dreaming and nondreaming sleep time.
7. If a mammal species experiences 5 hours of nondreaming sleep a day, how many hours of dreaming sleep would we expect that animal to get on average?
8. Calculate the difference in the number of hours spent in nondreaming and dreaming sleep for each mammal in the dataset.
9. What is the appropriate test to determine if mammals spend the same or different numbers of hours in dreaming and nondreaming sleep? Verify that the assumptions for using this test are met.
10. Write the hypotheses for the test you identified in Deliverable 9.
11. Conduct the hypothesis test and report your conclusion at the 0.05 significance level.
12. Create a 95% confidence interval for the mean difference in the number of hours a mammal spends in nondreaming and dreaming sleep.
13. Summarize your findings about dreaming and nondreaming sleep in mammals.

REFERENCE

1. Allison, T. and Cicchetti, D. (1976), "Sleep in Mammals: Ecological and Constitutional Correlates," *Science*, November 12, Vol. 194, pp. 732–734.

Index

H

I

J

L

M

N

O

P

Q

R

S

T

W

X

Z